Multidisciplinary Systems Engineering

James A. Crowder • John N. Carbone
Russell Demijohn

Multidisciplinary Systems Engineering

Architecting the Design Process

Springer

James A. Crowder
Chief Engineer, Raytheon Intelligence, Information and Services
E. Nichols Place, CO, USA

John N. Carbone
Engineering Fellow, Raytheon Intelligence, Information and Services
Garland, TX, USA

Russell Demijohn
Chief Engineer, Raytheon Intelligence, Information and Services
Aurora, CO, USA

Additional material to this book can be downloaded from http://springer.com

ISBN 978-3-319-22397-1 ISBN 978-3-319-22398-8 (eBook)
DOI 10.1007/978-3-319-22398-8

Library of Congress Control Number: 2015947764

Springer Cham Heidelberg New York Dordrecht London

Printed on acid-free paper

Springer International Publishing AG Switzerland is part of Springer Science+Business Media (www.springer.com)

Foreword

The authors, accomplished senior systems engineers, have developed a comprehensive book for systems engineering students by taking into account both theoretical and practical aspects of this important and growing discipline. In addition to covering all critical processes of systems engineering, they have uniquely provided a contextual introduction so that the reader learns about the systems engineering within the framework of all related disciplines. I see this as a unique contribution to the area and place this book at the top of the list.

At the very beginning of their introduction, the authors state the purpose of this book is "to arm the student with System Engineering principles, practices, and activities applicable to developing programs and systems within today's complex, distributed multi-discipline converging enterprise environments." They further promise to focus on "the overwhelming design gaps and needs of the current Systems Engineering discipline with foundations of new and relevant procedures, products, and implements."

The review of the content convinces me that the promise is delivered. Furthermore, they claim that the book is intended to be an introductory text book. In my opinion, although the book can be used as an introductory book, it can also be used in an advanced class with proper supplements. Another unique aspect of this book is its comprehensive multidisciplinary approach, which is not very common among systems engineering books. In this respect, the book is an answer to a 150-page roadmap published by the National Academy of the Sciences titled "Convergence: Facilitating Transdisciplinary Integration of Life Sciences, Physical Sciences, Engineering, and Beyond" (ISBN 978-0-309-30151-0). This comprehensive document states convergence as "an approach to problem solving that cuts across disciplinary boundaries. It integrates knowledge, tools, and ways of thinking from life and health sciences, physical, mathematical, and computational sciences, engineering disciplines, and beyond to form a comprehensive synthetic framework for tackling scientific and societal challenges that exist at the interfaces of multiple fields." Therefore, systems engineering, as an engineering of the twenty-first century, provides the engineering framework to the problems associated with convergence.

Interestingly, (SDPS) Society for Design and Process Science, www.sdpsnet.org, which the authors have been involved with from the beginning, has been investigating convergence issues since 1995. The founding technical principle of SDPS has been to identify the unique "approach to problem solving that cuts across disciplinary boundaries." The answer was the observation that the notions of Design and Process cut across all disciplines, and they should be studied scientifically in their own merits, while being applied for the engineering of artifacts. This book brings the design and process matters to the forefront of Systems Engineering, and as such, brings a complete flavor of convergence into the discipline of Systems Engineering.

During SDPS-2000 Keynote Speech, Nobel Laureate Herb Simon said, "... Today, complexity is a word that is much in fashion. We have learned very well that many of the systems that we are trying to deal with in our contemporary science and engineering are very complex indeed. They are so complex that it is not obvious that the powerful tricks and procedures that served us for four centuries or more in the development of modern science and engineering will enable us to understand and deal with them. We are learning that we need a science of complex systems, and we are beginning to construct it..."

This talk punctuated the initiation of engineering aspects in SDPS with the establishments of Software/Systems Engineering Society (SES) as a function of SDPS. Since then SDPS is transforming towards the next generation leadership. The current systems engineering book is a testimony that next generation leaders are producing. They are developing SDPS-SES into an organization that can address all the twenty-first century needs of software/systems professionals. There is not a single dedicated organization doing this nor are they equipped with the broad-based systems-oriented expertise that exists in SDPS. Furthermore, SDPS/SES should play a leading role in the development of the processes for next generation knowledge dissemination. The time for knowledge dissemination with classical methods and traditional journal publications is passing. We should play a critically leading role in this arena with new publishing initiatives. Book publishing with distinguished publishers such as Springer is one way of achieving this goal.

A decade later, in his SDPS-2010 keynote speech, Nobel Laureate Steven Weinberg described the hard realism of scientific knowledge generation combined with the key social role of scientists and engineers. These observations add to the urgency of our action as a society to spread "*thinking for understanding*" with the notions of Design and Process. Systems engineering books like this are a step in this direction.

SDPS is an international, cross-disciplinary, multicultural organization dedicated to transformative research and education through transdisciplinary means. SDPS is celebrating its 20th year in Dallas during November, 2015. Civilizations depend on technology and technology comes from knowledge. The integration of knowledge is the key for the twenty-first century problems.

This book is a very timely addition to serve this purpose and will be celebrated in our 20th year SDPS conference in Dallas. The comprehensive nature of the book, addressing complex twenty-first century engineering problems in a multidisciplinary manner, is something to be celebrated. I am, as one of the founders of SDPS, a military and commercial systems developer, and a teacher, very honored to write this foreword for this important book. I am convinced that it will serve generations to come in the growing arena of Systems Engineering.

Dr. Murat M. Tanik
Chair, Electrical and Computer Engineering Department
Wallace R. Bunn Endowed Chair for Telecommunications
UAB School of Engineering

Preface

The current global "content revolution" is characterized by information creation overload and mired in a vast swamp of ambiguous complexity and insecurity which we currently call "The Big Data" problem. Analogously, the Industrial Revolution spanned 50–100 years and was characterized by massive transition from hand production to machines. Each technology revolution has evolved from years of pooling knowledge until a critical mass of technology allowed for a major leap forward. We are swimming in oceans of data collectors and creators and drowning in the data. Simultaneously, we are trying to keep our head above water by spending billions of dollars to develop analytics to provide intelligence and knowledge from the massive data stores, as we try to automate the analysis, reasoning, and decision-making to handle our data problems. What is emerging is yet another great migration to handle the architecture and design of ever more complex System of Systems. The content revolution driven by seven billion people, five billion phones, one billion PCs, and ~90 PB of Facebook data holdings (2011) has resulted in current system designs that must fuse dozens of overlapping disciplines.

To further complicate current Systems Engineering efforts, there is growing interest in autonomous systems with cognitive skills to monitor, analyze, diagnose, and predict behaviors in real time that makes this problem even more challenging [139]. Systems today continue to struggle with satisfying the need to obtain actionable knowledge from an ever increasing and inherently duplicative store of non-context-specific, multidisciplinary information content. Hence, increased automation and complex System of Systems is the norm for current Systems Engineers and truly autonomous systems are the growing future. Additionally, the size, speed, and increased functionality of systems continue to increase rapidly, significantly challenging current Systems Engineering methods. Simultaneously however, development of valuable readily consumable knowledge and context quality continues to improve more slowly and incrementally.

Lastly, the complexity of systems and information today requires expertise in many disciplines and domains leaving engineering, just like during the industrial revolution, without the tools and level of understanding required to engineer across disciplines, much less obtain knowledge outside of their silos of expertise. Therefore, new Systems Engineering concepts, mechanisms, and implements are required to facilitate the development and competency of the Systems Engineering discipline and complex systems themselves, in order to simply be capable of proper operation, much less autonomous operation, self-healing, and critical self-management of knowledge and real-time operational self-awareness. Presented in this first of a series of books are new Multidisciplinary Systems Engineering concepts, processes, methodologies, notional architectures, and tools to support the engineering discipline in understanding and evolving engineering of systems across the full spectrum of disciplines required today. The materials include the rationale for Multidisciplinary System Engineering (MDSE) as the standard for systems engineering and development; ensuring new System of Systems (SoS) designs are successful and avoid the failures introduced by complexities of the information overloads facing our customer and management teams.

E. Nichols Place, CO — James A. Crowder
Garland, TX — John N. Carbone
Aurora, CO — Russell P. Demijohn

Contents

List of Figures

List of Tables

Chapter 1
Introduction: Systems Engineering—Why?

The purpose of this text book is to arm the student with System Engineering principles, practices, and activities applicable to developing programs and systems within today's complex, distributed multi-discipline converging enterprise environments. Specifically, the focus is to match the overwhelming design gaps and needs of the current Systems Engineering discipline with foundations of new and relevant procedures, products and implements. Therefore, this introductory text book provides the basis for a modern Multi-disciplinary Systems Engineering approach.

1.1 The Need for Formal Systems Engineering

System Engineering is an overarching process that trades off and integrates capabilities within a system's design to achieve the best overall product and/or capabilities. To achieve successful design solutions, System Engineering requires quantitative and qualitative decision making. This involves trade studies, optimization, selection, and integration of the results from many engineering disciplines [1]. This is accomplished via an iterative process that derives and defines requirements at each level of the system, beginning at the top level requirements and propagates those requirements through a series of steps that eventually leads to a physical design at all levels. Iteration and design refinement lead successively to preliminary design, detailed design, and final approved design. At each successive level, there are supporting lower level design iterations that are necessary to gain confidence in the design decisions. During these iterations, many conceptual alternatives are proposed, analyzed, and evaluated in trade studies, resulting in a multi-tier set of requirements. These requirements form the basis for controlled, formal verification of performance. System Engineering (SE) closely monitors all development activities and integrates the results to provide the best solution at all levels of the system.

J.A. Crowder et al., *Multidisciplinary Systems Engineering*,
DOI 10.1007/978-3-319-22398-8_1

One of the most important reasons SE exists is that it provides the context, discipline, and tools to adequately identify, refine, and manage all system requirements in a balanced manner. Systems engineering provides the disciplines required to produce comprehensive solution concepts and system architectures. It also provides the discipline and tools to ensure that the resulting system meets all requirements that are feasible within specified constraints.

Uncertainty and Risk are intrinsic components of engineering projects. One of the major challenges of System Engineering is the effective management of performance, cost, schedule, technology, and risks. Most of all, Systems Engineers are held responsible for ensuring that the system doesn't simply perform to the letter of the requirements but that it performs to the expectations and the critical needs of the system user. Therefore, any student of Systems Engineering must understand that true success in Systems Engineering will include Science and Art across many disciplines. System Engineering has had successes, as well as failures. Some of the major lessons learned are:

1. Formal System Engineering processes are essential, but not sufficient for good System Engineering implementation.
2. Successful System Engineering requires:
   (a) Systems-level thinking: Having the discipline to look at the entire system and perform functional analysis and decomposition without thinking about implementation. Understanding how our systems fit within the context of the overall enterprise and the specifics of our customer's mission simultaneously.
   (b) Sound engineering development processes and procedures.
   (c) Proven risk management processes

1.2 A Brief Historical Perspective

Systems engineering began its development as a formal discipline during the 40s and 50s at Bell Laboratories. It was further refined and formalized during the 1960s during the NASA Apollo program. Given the aggressive schedule of the Apollo program, NASA recognized that a formal methodology of systems engineering was needed, allowing each subsystem across the Apollo project to be integrated into a whole system, even though it was composed of hundreds of diverse, specialized structures, sub-functions and components. The discipline of system engineering allows designers to deal with a system that has multiple objectives with a balance that must be struck between objectives which differ wildly from system to system. System engineering seeks to optimize the overall system functionality, utilizing weighted objectives and trade-offs in order to achieve overall system compatibility and functionality [2]. During the 70s and 80s as engineering systems continued to increase in complexity, it became increasingly difficult to design each new system with a blank page. As system quality attributes like reliability, maintainability, re-usability, availability, etc. became more and more important, the concept of Object-Oriented design techniques were developed. The first Object-Oriented languages began to emerge during the 70s. By the 1980s, the first books on Object-Oriented Analysis and Design (OOAD) were published and available. Unfortunately there were many different OOAD methodologies and no consistency among methods. At one point in the 90s, there were over 50 different OOAD methods [3]. This became increasingly difficult for the Department of Defense (DoD), because contract proposals from competing contractors utilized entirely different OOAD methods to design their systems, making comparison between proposals nearly impossible. Finally, in 1993, the Rational Software Company began the development of a Unified Modeling Language (UML), based on methodologies by Grady Booch, James Rumbaugh, Ivar Jacobsen, coupled with elements of other methods [4]. Here the Rational Software Company simplified current methods from several authors into a set of OOAD methods that included Class Diagrams, Use Case Diagrams, State Diagrams, Activity Diagrams, Data Flow Diagrams, and many others [5].

1.3 Systems Engineering Practices and Principles

As mentioned earlier, System Engineering requires both art and science for successful design, implementation, test, sell-off, and acceptance of programs. Therefore, successful programs require that Systems Engineering embody the following principles [6]:

1. Tailor the SE activities to the scope and complexity of the program.
2. Ensure that the system design meets the needs of the customer and addresses the complete life-cycle for the system.
3. Acts as the interface and "glue" between the other engineering disciplines to ensure that:
   (a) The hardware and software components of the system meet their requirements, functionality, and are operationally supportable.
   (b) Subsystems are compatible with each other.
4. Establish and maintain the System Architecture
5. Ensure that the system is compatible with all external interfaces.
6. Establish and maintain requirements. This includes planning for requirement changes as insight into the need and the program solution and implementation "evolves."
7. Manage technology and innovation by generating a wide range of implementation alternatives before converging on a solution (trade studies).
8. Understand the program risk/benefit trade-off strategies among performance, cost, and schedule.
9. Manage technical risks and opportunities for the program.
10. Manage and maintain quality throughout the program.
11. Create and maintain appropriate system documentation.
12. Institute continuous improvement.
13. Overall oversight of all technical activities.

Along with these System Engineering activities, the System Architecture and three overarching processes that interact with all program/system activities throughout all engineering disciplines are employed. These processes, which continue throughout the system's Life Cycle, are described below: Integrated Technical Planning, Risk Management and Integrity of Analyses, and System Architecture.

1.3.1 Integrated Technical Planning

Integrated Technical Planning provides the technical guidance tools required to track and manage program activities at every level within the system design and implementation. The goal of Integrated Technical Planning is to ensure cooperation and participation from all customers. This includes the management, and engineering staff in order to examine and agree on the economic, social, environmental costs and overall benefits of the system and system design. This allows the project MDSEs to choose the most appropriate design options, and to plan viable and suitable courses of action throughout the system life cycle. This includes the Integrated Master Plan and Integrated Master Schedule for the system:

- Integrated Master Plan (IMP): an event-based plan providing a hierarchical view of project/system events; each being supported by specific tasks with entry and exit criteria for each. The IMP usually provides a narrative which explains the overall management philosophy for the project.
- Integrated Master Schedule (IMS): an integrated and networked schedule that contains all the project work and planning packages required to support the events, accomplishments, and criteria laid out in the Integrated Master Plan. In short, the IMS should be directly traceable to the IMP, where the schedule tasks have measurable entry and exit criteria to ensure IMP criteria satisfaction. This includes all elements required to drive the development, implementation, testing, production, and delivery of the system.

1.3.2 Risk Management

The Role of Risk Management is to provide an organized, systematic decision-making approach to identify risks that affect achievement of program goals. Risk management involves the analysis, assessment, control, avoidance, minimization, and/or elimination of unacceptable risks within the system and overall project. This includes the identification, assessment, and prioritization of potential risks to the technical, schedule, and cost across the entire project/system life cycle. The purpose of Risk Management is to control the probability and/or impact of risk events to the program, and to maximize possible opportunities that benefit the program. Risk Engineering is discussed in detail later in this chapter and later in the book.

1.3.3 Integrity of Analyses (Integrity Engineering)

Integrity Engineering ensures provision of credible, useful, and sufficient data/results for program management's decisions-making process throughout the program's Life Cycle and ensures the integrity and fidelity of the various analysis processes, trade studies and tools. Integrity Engineering's involvement throughout the engineering Life Cycle helps to ensure that the end product, the engineering processes, and the overall system design meets the appropriate and intended requirements. Integrity Engineering seeks to acquire the technical, economic, social, legal, and practical knowledge required to provide assurance and verification of functionality to ensure that the system/project meets and continues to meet the mission/business requirements, as well as the safety, legal, performance, and other requirements for overall system long-term viability.

1.3.4 System Architecture

The system architecture is a conceptual, physical and logical blueprint that defines the structure and operation of a system or system of systems. The intent of the systems architecture is to determine what the system does, and how a system can most effectively achieve current and future objectives [7]. The system architecture is utilized within a project's organization to integrate people, technology, and information resources in sometimes vastly different proportions, based upon the overall goals and Conceptual Operations (CONOPS) for the system. As a result, all system architectures are unique. Each has unique requirements, depending on the overall business or mission goals and Strategic Vision(s) of the overall project. Hence, the result can comprise situations where Revenues and Profits scale differently, depending on the required labor, technology, software, and information needs. Figure 1.1 illustrates this concept.

The discipline of Systems Architecture and especially of System of Systems Enterprise Architecture is a multi-tier, multi-discipline engineering specialty that requires an understanding of the strategic, mission/business and technical requirements, their processes, and their interrelationships [8]. Figure 1.2 illustrates the layers and functionality of the Enterprise Systems Architecture process.

Several architecture types are required within a complete System of Systems Enterprise Architecture:

- **Business/Mission Architecture**: defines the mission/business processes and strategies required, based upon the CONOPS and System Requirements. Business/Mission architecture takes into account the processes required for system development and operations, as well as the goals, objectives, technology environment, and external interfaces. Business/Mission process mapping allows these processes to be translated into an overall Enterprise Information Strategy. These information strategies are utilized to define the overall Enterprise Information Architecture.

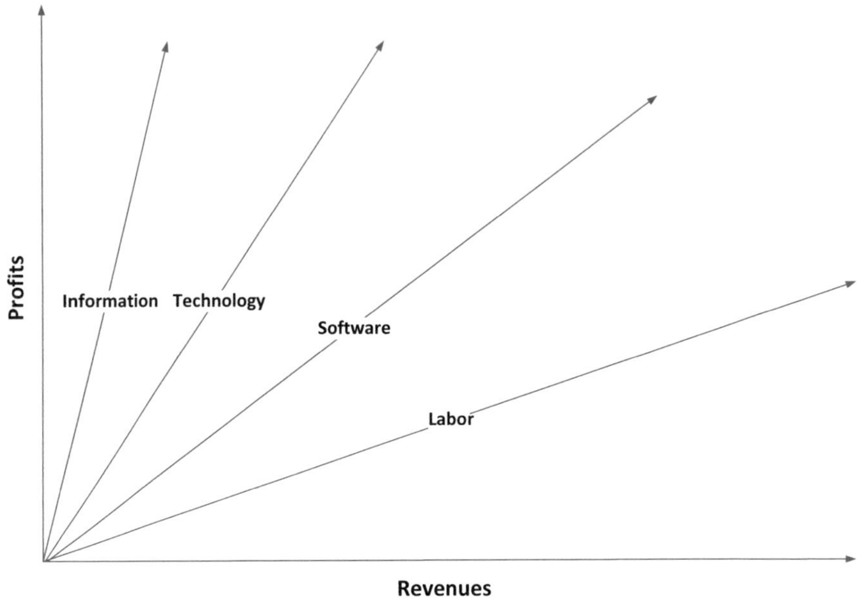

Fig. 1.1 Revenue vs. profits scalability

- **Information Architecture**: defines the information required for the application-level aspects of the system and how the information flows through each of the high-level system functions. This includes all Business/Mission processes, the external interfaces, data flows, user interfaces; mapping information needs to system functionality. The Information Architecture may be constrained by limitations present at the Business/Mission process level (e.g., viability of specific applications because of bandwidth constraints).
- **Data Architecture**: this architecture flows from the Information and Business/Mission Architectures and its design derives from the need to make decisions about how data will serve the system as part of each Business/Mission process and information usage across the Enterprise Architecture. The Data Architecture defines the current and future needs for accumulation, usage, renewal, maintenance, and transfer controls of data within the system enterprise architecture, as well as outside the system's boundaries (external interface data needs). Data Architecture aligns the many data related system aspects with the Business/Mission applications, system related protocols and all hardware and software applications. From the Enterprise Architecture perspective, the Data Architecture must consider volume, velocity, variety and veracity, issues, such as:
    - Repositories to facilitate storage and retrieval of system information.
    - Data mining for gathering and correlating data.
    - Data formats and protocols for integration of the systems mission/business processes.
    - Data integrity and security (e.g. internal, external).

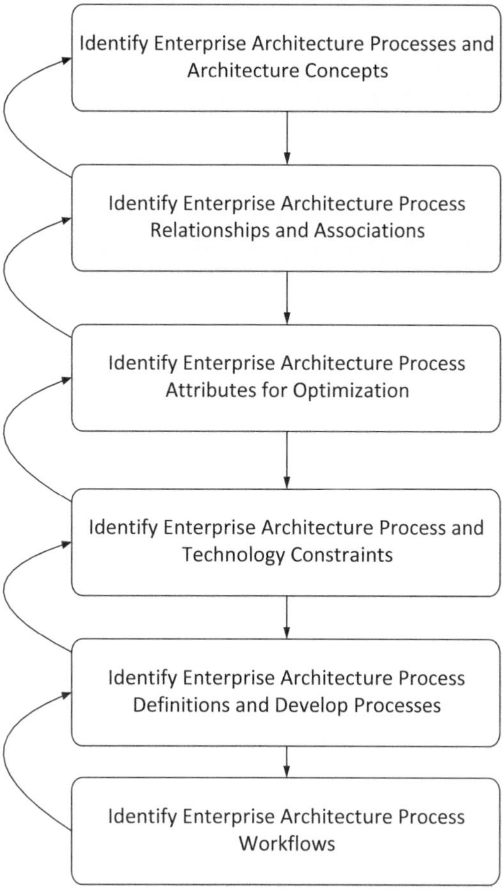

Fig. 1.2 Enterprise architecture engineering process flow

  - Data provenance.
  - Data responsibility, stewardship and ownership.
  - Data internal and external performance.
  - Data warehousing that considers the system's data requirements and accessibility (e.g. Near Line, Off-Line).
  - Data modeling tools.
  - Development tools.
  - Data dictionaries and query languages.

- **Communications Architecture**: this relates the Information Architecture, the Data Architecture, and the Computer/Network Architecture to how data is communicated. Details about the specifics of the Communications Architecture define how the system is going to meet the demands made by the mission/business processes, the data requirements, as well as the hardware and software that will be required to support them. Examples include Client/Server Architectures, messaging requirements, data flow requirements, and transaction management.

- **Computer Architecture**: this relates the Data Architecture, Information Architecture, and Systems Architecture to how these architectures will be physical and logical instantiated. This primarily consists of specific hardware and software, often commercial, off-the-shelf (COTS) hardware and software that constitute the technological base for the above architectures. Product availability and budget allocations may affects and determine the choice of specific hardware and software for the system.
- **Software Architecture**: refers to the high-level structure of the software system. Important properties of the software architecture include [9]:
  - The software architecture is initially defined at a high enough level of abstraction such that the system can be viewed as a whole across the whole system enterprise.
  - The software structure must support the mission/business processes of the system. This drives the software architect to take into account the dynamic behavior of system in order to ensure the Software Architecture will support all current uses and as many as possible future uses of the system.
  - The Software Architecture must take into account functional as well as non-functional requirements (called Quality Attributes) for the system [10]. Quality Attributes characterize the general overarching quality of the system which can include, but are not limited to, performance, security, availability, and reliability. In addition, the Software Architecture must comprise flexibility and extensibility requirements associated with accommodating the mobility and pliable nature of the system design under construction and the future extensibility and viability of the functionality being designed into the to-be system. These requirements can conflict and therefore tradeoffs must be continuously assessed and alternatives considered during the design [11]. As an example, the more discrete or modular that the functional components of the software are designed the more pliable it will be. However, this can affect performance in terms of the number of context switches that can potentially take place and the potential amount of effort the system has to employ to perform menial tasks. Hence, the proper system balance must be achieved.

1.4 Inter-discipline Relationships

System Engineering activities are an integral part of a program's life-cycle. They complement the program management and the design activities by placing greater emphasis on iterative development, trade studies, uncertainties, and technical risk management in order to optimize program success, to include providing and analyzing Technical Performance measures to ensure performance within cost and schedule constraints.

As we move from classical systems engineering to more viable modern System of Systems Engineering (including agile Systems Engineering) approaches, the rapidly converging global environment is driving us to comprehend and employ increasing numbers of disciplines in the overall architectural design process [12].

1.4.1 Multidisciplinary Systems Engineering

Notions of inter-multi-disciplinary approaches to problem solving are not new. From the development of opposing thumbs the human race has continued to automate and optimize. Initially driven by survival, whether preparing for the onset of deep winter, or more efficiently hunting for food, or finding a way to cave paint at night, understanding the environment and everything in it was absolutely fundamental. Today, this remains unchanged regardless of the advanced scientific methods and implements at our disposal. What is true is that the further we forge, enhance, automate, and improve, the greater complexities we encounter. We vigorously attack each new challenging complexity with our centuries of powerful tricks and procedures [13]. We consistently discover new needs for developing a new type of hammer or nail, or a hammer which simultaneously drives many nails at once. Thus far, our survival has been predicated by the understanding of our environment, creating language, telling stories and the writing down, whether on papyrus or paper or digital files, the continuously learned fundamental truths.

These fundamental truths became our driving principles, increasingly codified over time. A thousand years ago, the root word of "Principle", "Prin-" or "primus", meant a class preeminent person or thing, and "-ciple" derived from "cep", meant "to take" [14]. In summary, the word, "Principle" comprises the taking-on of intrinsic value, and high importance as the authors of the Declaration of Independence, Newton, Spencer, Euclid, or Aristotle himself must have felt while developing the intrinsic value within their disciplines, and the inspiration and foundational understanding students receive. Subsequently, we began to manage, organize and expand our evolving "First Principles" into topical areas we know today as ***Disciplines***. Figure 1.3, depicts Discipline influences, essential characteristics and evolution of a Discipline's knowledge base accomplished via natural human means: mental insights, scientific inquiry process, sensing, actions, and experiences [15].

This approach served us very well except that most disciplines evolved over centuries as inward looking areas of specific and valuable myopic expertise. Hence, as information content and science expands at "warp" speed and engineering continues to explode with complex interdependencies, disciplines become wrought with new high levels of uncertainty and many new unknowns. Therefore, to achieve discipline advancement we are driven to broaden our scope of investigation beyond the boundaries of the existing discipline's First principles and concepts. This means stretching the boundaries of knowledge by traditional observation and analysis to our comprehensive, albeit myopic, disciplinary environment, falling back to our survival instincts and natural cave man tendencies. Problem is that to do so requires modern mechanisms which do not exist in engineering disciplines today to match the vast modern complexities required for engineering solutions across the comprehensive environment of many disciplines. This requires education, research and new methods that can transcend traditional disciplinary or organizational boundaries, to enable the solution of large complex problems by teams of people from diverse backgrounds [16].

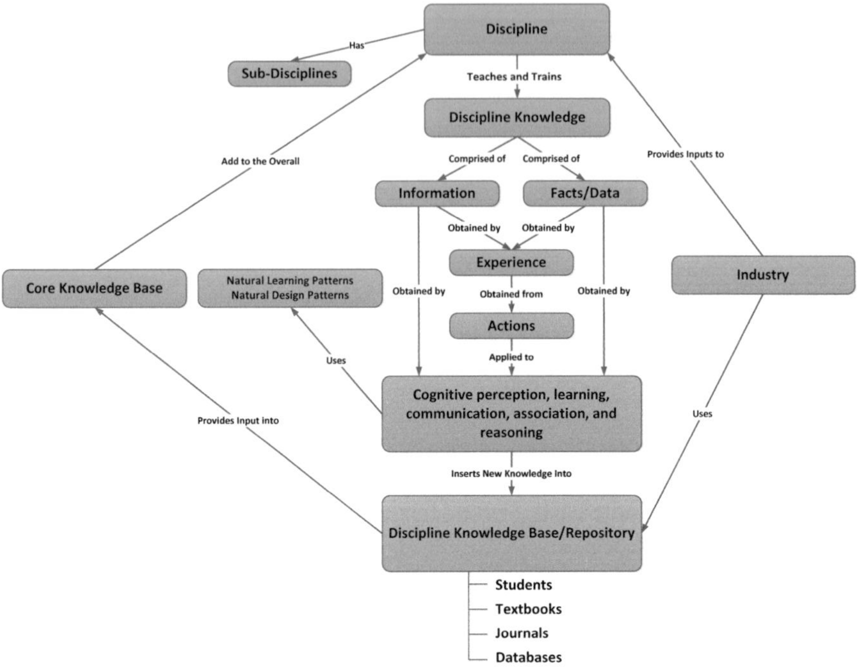

Fig. 1.3 Discipline evolution

Systems Engineering is, by its nature, a multidisciplinary science, endeavoring to look at the "big picture", defining what a system must do; encompassing all aspects, including functionality, performance, reliability, etc. [17]. Figure 1.4, describe Multi-disciplinary approach as one where *methods* from two or more disciplines are examined to determine topic benefits within any one discipline whereas, Inter-Disciplinary work involves the transfer of beneficial method(s) applied to topics between disciplines. Both approaches cross discipline boundaries but still remain within the boundary of the independent researchers found within one given discipline; hence, minimal benefits are realized [18].

However, significant opportunities still remain uncharted for promoting deeper computational type thinking and understanding, analysis, cooperation between disciplines to achieve more compelling and significant solutions to the complex problems we face today. This is known as the Transdisciplinary approach. Researchers from diverse disciplines work jointly to develop the necessary implements, and shared conceptual frameworks through the use of computational thinking by creating different levels of abstraction to understand and solve problems more effectively.

The development of Computer Science discipline, over the last three decades, has created an inherent revolutionary way of thinking and approaching problems due to the vast application of computational needs across every scientific domain. True research cannot be accomplished today without Computational Thinking; Systems Engineering being responsible for handling the overarching delivery of any project or program is sorely lacking this expertise. Additionally, the systems

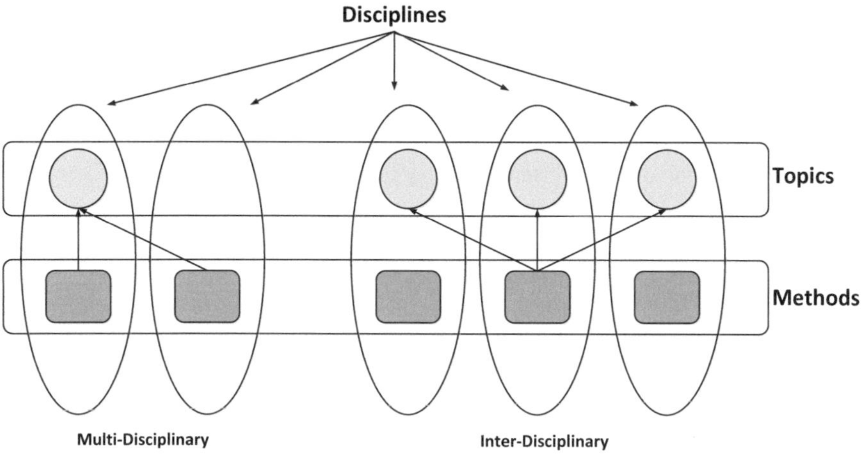

Fig. 1.4 Discipline type comparison

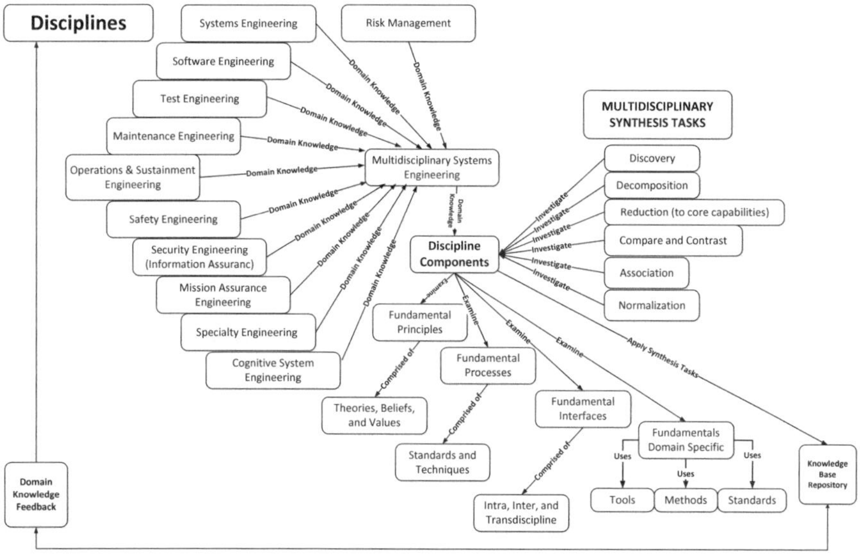

Fig. 1.5 Multidisciplinary systems engineering process flow

engineers working in various domains have little to no understanding, theory, tools, concepts, methods or guidance to assist them in engineering across the many disciplines they are responsible to understand in order to develop viable solutions in the vastly complex environments they are thrust into today. The following chapters describe concepts, tools and methodologies and their application for Systems Engineers to develop usable Transdisciplinary applications to support Multidisciplinary and Interdisciplinary efforts. Figure 1.5 illustrates this multidisciplinary approach for Systems Engineering.

Figure 1.5 illustrates many disciplines that are required for modern, Systems of Systems engineering designs. Especially future systems that will require machine learning, reasoning, and autonomous decision making capabilities. Each of these disciplines brings necessary perspectives to the overall Systems Engineering requirements derivation and architectural designs. Definitions of each discipline and their ties to the overall Systems Engineering discipline are discussed below:

1.4.2 Systems Engineering Process

Systems Engineering: a methodical and disciplined process to define requirements, system design, realization, technical management, operations, and retirement of engineering systems. Figure 1.6 below illustrates the high-level Systems Engineering Process.

1.4.3 Software Engineering Process

In 1968 and 1969, the NATO Science Committee sponsored two conferences on software engineering, seen by many as the official start of formal discipline of Software Engineering. The discipline grew, based on what has been deemed the "Software Crisis" of the late 60s, 70s and 80s, in which very many major software projects ran over budget and over schedule; many even caused loss of life and property. Part of

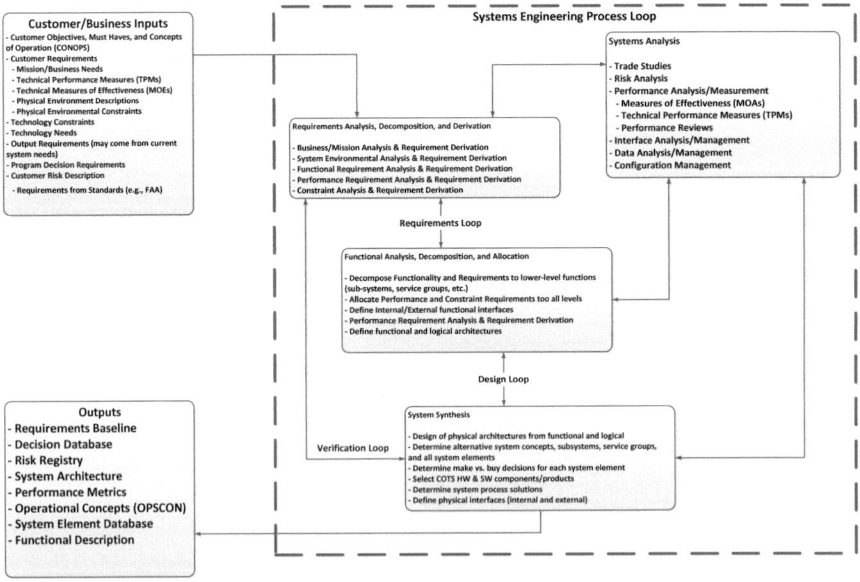

Fig. 1.6 Basic systems engineering process

the issues involved in software engineering efforts throughout the 70s and 80s is that they emphasized productivity, often at the cost of quality [19]. Fred Books, in the Mythical Man Month [20] admits that he made a multi-million dollar mistake by not completing the architecture before beginning software development; a major problem that has been repeated over and over, even today. We will discuss the notion of the importance of having a complete architecture later in the book. This does not mean that the architecture can't change, as it often does throughout the project, but both the system and software engineering process must have proper feedback in order keep up with changes so that the development teams clearly understand the architecture they are developing to during every sprint [21]. We will discuss this at length in subsequent chapters, as the intent here is just to provide a brief history.

There are many Software Engineering methodologies that have been developed through the years. These include waterfall, Iterative, Spiral, Agile, and Extreme software development cycles, each with their own set of advantages and disadvantages. However, for any given development approach, there are basic steps that must be taken. Here we will assume a more feedback-driven, agile approach, understanding that the requirements may change at any time during the development process, and therefore changes throughout the system may need to be made to accommodate the agile nature of modern system design and implementation. Figure 1.7 illustrates this Software Engineering Process.

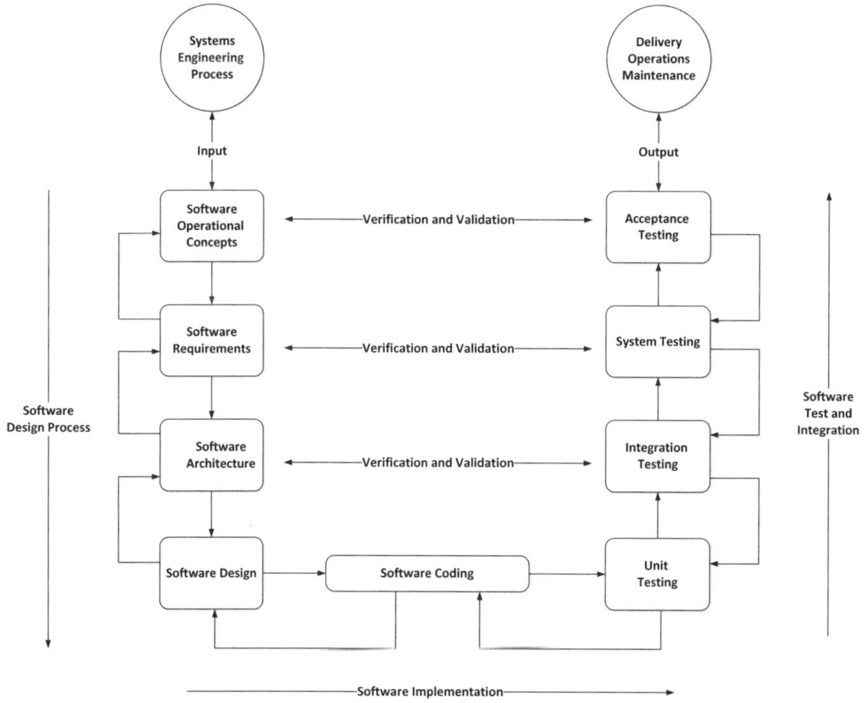

Fig. 1.7 Basic software engineering process

Here we specifically do not show this process as a "V," a "Spiral," or any other standard development diagram, since the multidisciplinary approach to system development should be utilized with any standard or non-standard development methodology. Figure 1.7 does demonstrate the feedback-driven nature of true multidisciplinary engineering, where each discipline is intimately involved in all steps in the systems, software, testing, and delivery of engineering systems.

1.4.4 Test Engineering Process

Test Engineering is responsible for determining how to test the component of the system to include the finished system such that it produces 100 % coverage of all technical, performance, non-technical, and quality requirements (e.g., Reliability, Maintainability, and Availability). Test engineering should be (but often is not) included in the early stages of the design process. This helps to ensure the system design includes testability, maintainability, and manufacturability. In short, the test engineering process ensures that the capabilities can be built and readily tested. Too often companies and projects look to save money by taking shortcuts with the test engineering process. In every case, when the test engineering process is bypassed, testability becomes overly complicated later in the implementation process causing bottlenecks and delays in the development, testing, delivery, and maintenance of the overall system. Figure 1.8 illustrates this test analysis/design/implementation process, which builds on the Software Development Process from Fig. 1.8.

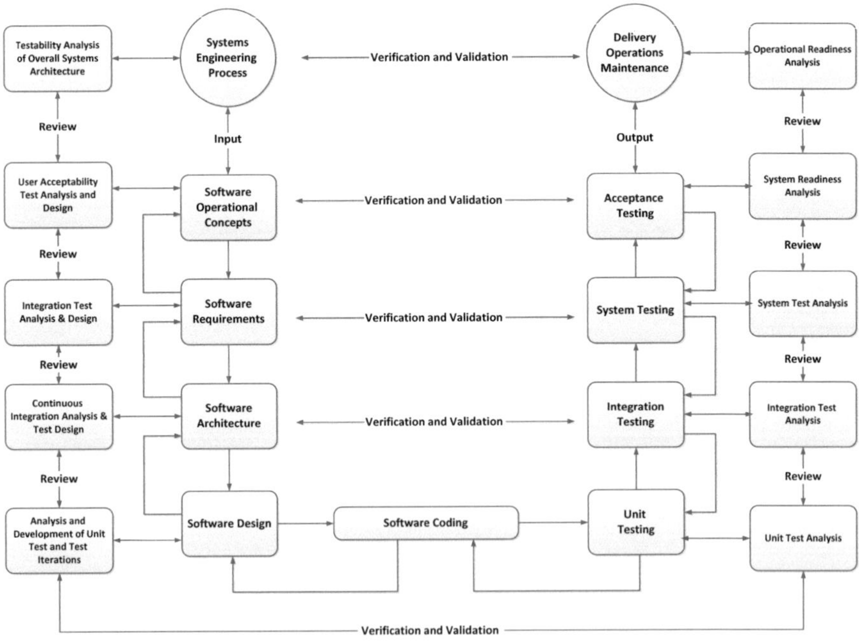

Fig. 1.8 Basic integrated test engineering process

1.4.5 Maintenance Engineering Process

The Maintenance Engineering Process is an approach to influence the maintenance activities through the design of the system, both hardware and software. The output of this process provides a set of guidelines which serve as a tool to apply Design for Maintenance to the system design. The guidelines describe various ways how these future maintenance activities can be influenced. Applying the guidelines can help to reduce the number of maintenance activities, to make them easier to execute or to decrease the logistic support time that is required for them.

1.4.6 Operations and Sustainment Engineering Process

The Operations and Sustainment Engineering Process is intended to provide an understanding of operational and sustainable concepts throughout the design, implementation, test, delivery, and operations phases of a project (i.e., the entire life cycle). It allows the design and implementation teams to understand how end users will utilize the new or improved systems, including how their jobs will change as a result of this development effort. Without Operations and Sustainment as part of the overall multidisciplinary engineering process, the systems, software, and test engineers typically will not take usability into account in the requirements, architecture, development, or testing of the product/system. This can radically decrease the user buy in of the new system, which can result in the new/improved system not being used or possibly even sabotaged. Operations and Sustainment must be part of an overall lifecycle analysis throughout the entire engineering processes. Sustainability concerns may be dealt with as constraints on the overall design of the system.

1.4.7 Safety Engineering Process

System Safety Engineering is typically applied to complex and critical systems (e.g., national power grid, DoD weapon systems, etc.). The methodologies and tools utilized by safety engineering are there to present, eliminate, and control hazards and overall safety concerns [22]. Without including safety engineering as an integral part of the overall Systems Engineering process, designs may have to be redone, or scrapped all together late in the development process. Software Safety Engineering is an emerging field applied to modern systems that seeks to ensure that safety-critical systems that are under software control are designed to mitigate risk to acceptable levels. Again, without software safety engineering as an integral part of an overall multidisciplinary engineering process, entire software designs may have to be redone and recoded late in the development process, greatly increases the cost and schedule of the system/projects.

1.4.8 *Security Engineering Process*

The Security Engineering discipline endeavors to develop Information Assurance solutions, which are applied to the system architecture, to protect both the system and its information/data for both isolated and network connected systems. The Security Engineering process is also applied to the Systems of Systems Enterprise Architecture to ensure system data/information maintains:

- Integrity: data/information remains unchanged as it is flowed through the system and has not been "tampered with."
- Confidentiality: data/information is available only to those who have the need and credentials to access the information, whether this is a person or a process.
- Authentication: this ensures that users/processes are who they are supposed to be. This includes the use of passwords, digital tokens, biometrics, or other methods.
- Availability: ensures data/information is available for use by the users/processes that need it. This includes protection from hackers or processes that would block access to information or information systems critical to business/mission needs.
- Nonrepudiation: creates proof that a user/process completed an action on the system and cannot deny having access to or using the system.

The Security Engineering process assesses the system for potential threats and creates security (information assurance) architectures with appropriate safeguards in order to maintain the system's operational integrity and long-term viability. The goal of Security Engineering is to keep long-term operations within acceptable mission/business integrity boundaries with manageable risks. Figure 1.9 illustrates

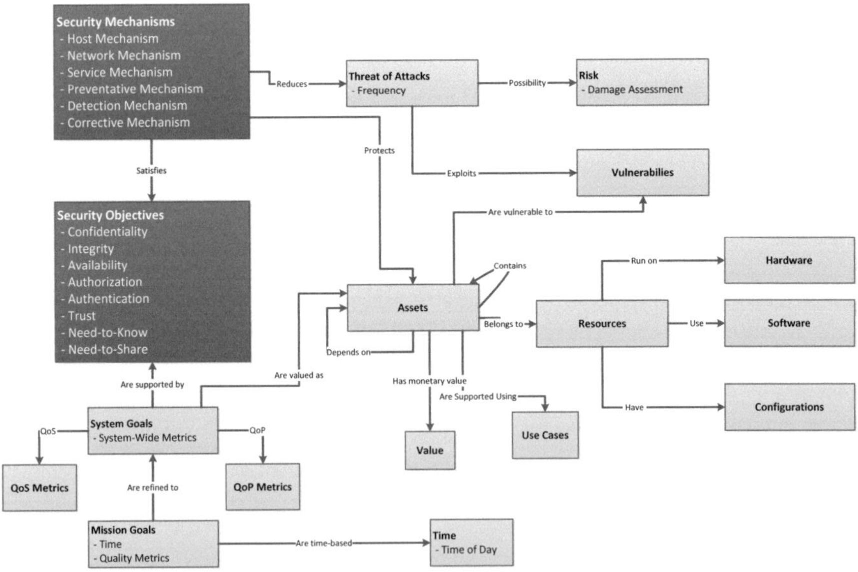

Fig. 1.9 Basic security engineering process

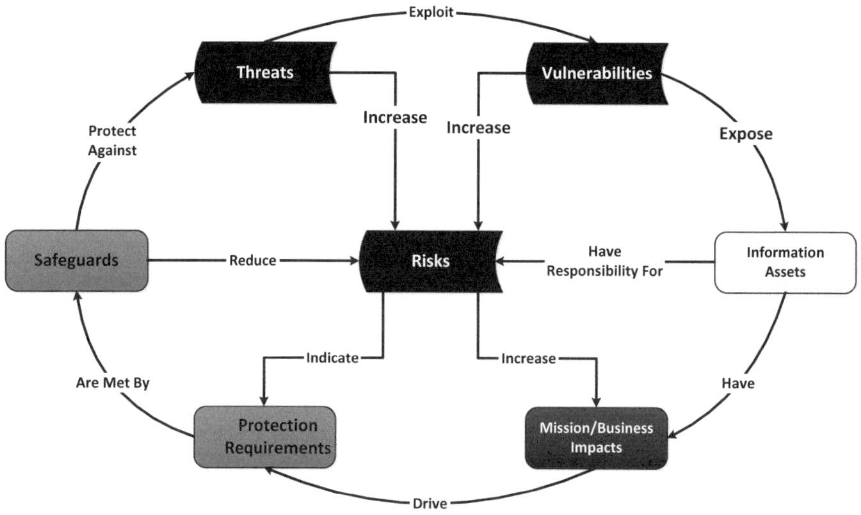

Fig. 1.10 Security engineering assessment process

the high-level Security Engineering Process. Figure 1.10 depicts the Security Engineering assessment process that looks at the system risks, threats, and vulnerabilities and how to protect the system from realizing these.

1.4.9 Mission Assurance Engineering Process

The role of Mission Assurance is to identify and mitigate deficiencies within the system, in terms of:

- Design
- Production
- Test
- System fielding/installation
- Operations

The goal of Mission Assurance is 100 % customer satisfaction. Mission Assurance is a full life-cycle process that should be involved in every aspect of the system, from conceptual design to end-of-life termination of the system. Mission Assurance Engineering includes Systems Engineering, Risk Engineering, Quality Engineering, and Management practices and principles to achieve mission/business success. Mission Assurance Engineering emphasizes system survivability, which may involve deriving requirements for redundant backup capabilities, engineering failover modes of operation, and roll-back capabilities for system upgrades and modernization. Mission Assurance Engineering may involve assessment of potential adverse actions

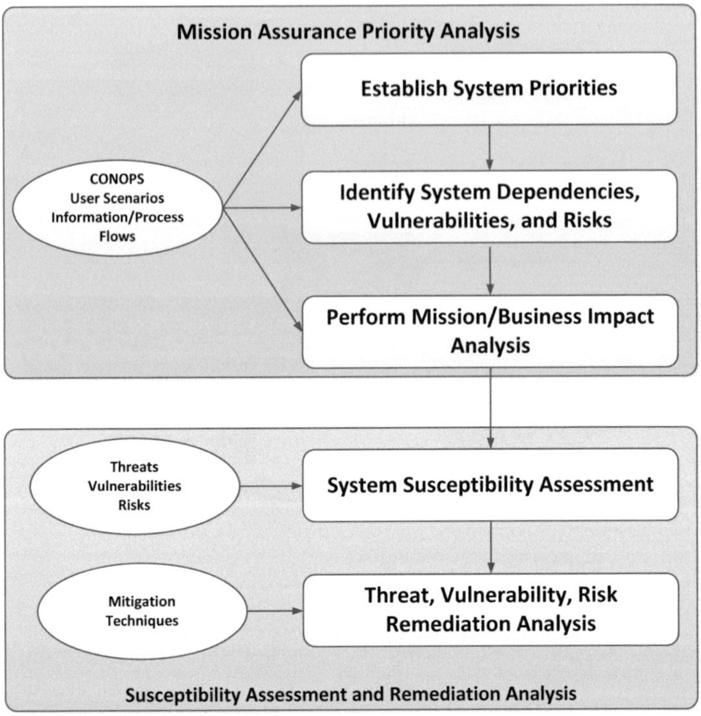

Fig. 1.11 Basic mission assurance engineering process

and ensure the system architecture can operate under these conditions. The inclusion of the Mission Assurance Engineering processes help to ensure that the systems are:

- Usable
- Dependable
- Robustness
- Agility
- Flexibility
- Adaptability
- Resiliency

Figure 1.11 illustrates the Mission Assurance Engineering Process.

1.4.10 Specialty Engineering Process

While Systems, Software, and Test Engineering play the primary roles in the overall system design and implementation, Specialty Engineering encompasses as host of specialty disciplines that are crucial to the overall success of any system development effort. From the very beginning of conceptual analysis where Specialty Engineering is involved in areas that include, but not limited to:

- Structural Integrity
- Aerodynamics
- Thermodynamics

Specialty Engineering is involved in the initial design process with such disciplines as:

- Affordability
- Quality Assurance
- Reliability
- Producibility
- Logistics
- Human Factors

These Specialty Engineering disciplines drive constraints on requirements derivation to ensure viability of the overall system design. Once the draft design is worked out, the Specialty Engineering process brings another set of specialty disciplines into play in order to refine the final requirements; further detailing the system design and helping to shape the detailed development plans [23]:

- Facilities Design
- Equipment Design
- Procedural Requirements
- Personnel Requirements

An example of Specialty Engineering is shown in Fig. 1.12, which illustrates the Human Factors Engineering Process.

1.4.11 Cognitive Systems Engineering Process

Cognitive Systems Engineering is a fairly new engineering discipline that draws on insights from cognitive, social, and organizational psychology to address the social-technical aspects of modern system design. Cognitive System Engineering seeks to enhance or amplify human capabilities to perform cognitive work during the operation of the system (e.g., UAV operators). The intent of Cognitive Systems Engineering is to integrate the technical function of the overall system needs with human capabilities (human cognitive processes) to operate the system and to make that cognitive work more reliable. This includes cognitive functionality relating to decision management, planning, collaborating, and system management. Cognitive Systems Engineering works with Human Factors Specialty Engineering in the design of human-machine interfaces, human-human and human-machine communications, technology training systems, and management teams.

The rapid rise of computer-based technologies over the last few decades has driven the need for Cognitive Systems Engineering to provide a comprehensive

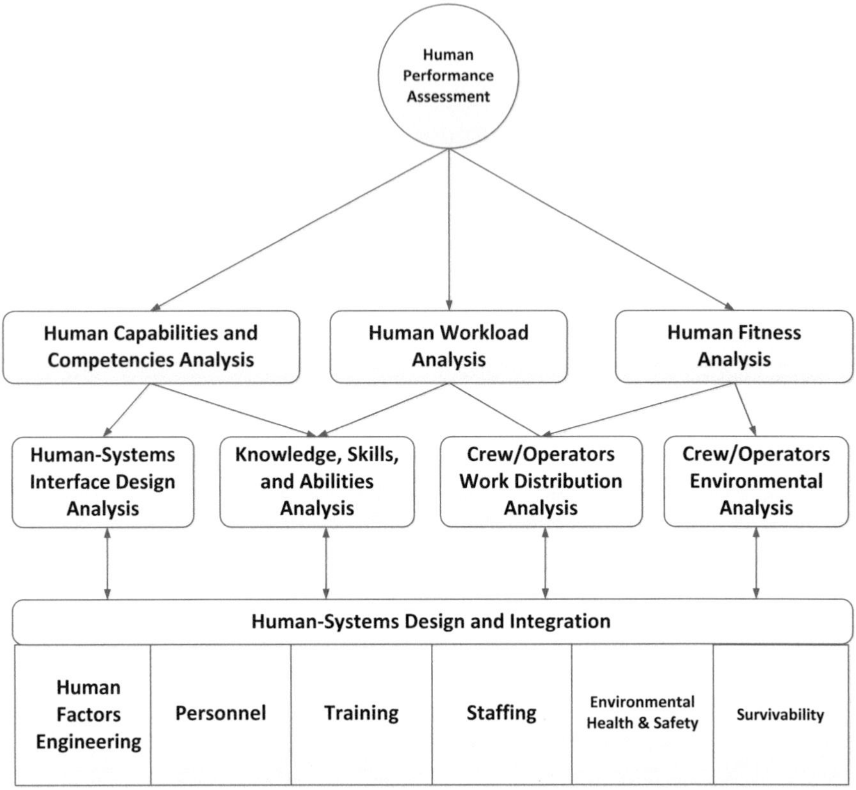

Fig. 1.12 Human factors engineering process

approach for the design of modern social-technical systems. The design, development, and implementation of complex dynamic information environments and decision intensive systems (e.g., air traffic control) has created the need for an engineering discipline that takes into account human cognitive states, human cognitive processes, and human cognitive strategies in the overall systems design. Cognitive Systems Engineering seeks to develop and embed methods and constructs into the overall systems design that provide decision and planning tools that support and enhance human cognition in the operation of the system. Figure 1.13 illustrates the Cognitive System Engineering Process.

1.4.12 Risk Management

Risk, as applied to SoS development, is the measure of a potential inability to achieve the overall program objectives within the defined cost, schedule, and technical constraints provided. Risk has two components, the likelihood that a risk event that

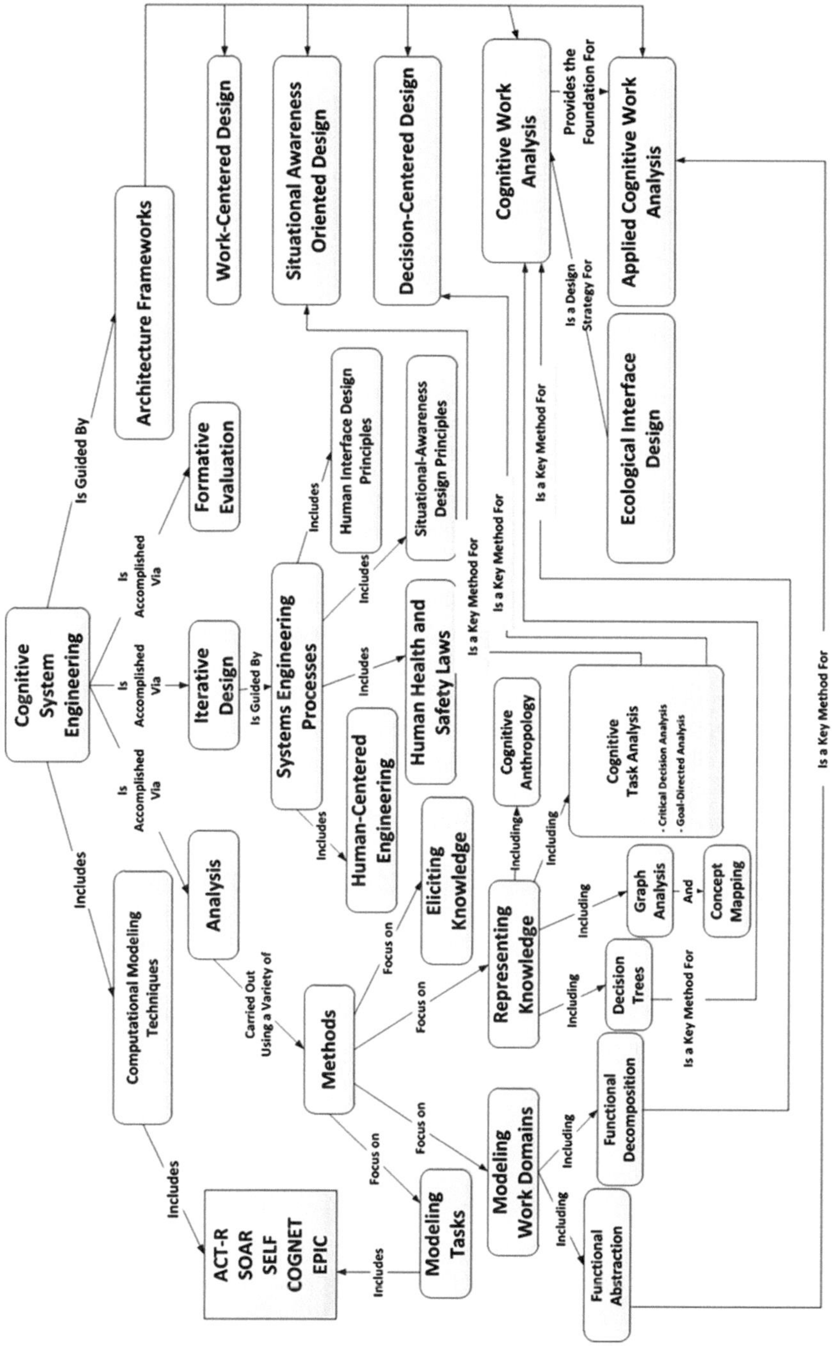

Fig. 1.13 Basic cognitive system engineering process

Fig. 1.14 The risk management recurring process

would prevent the program from achieving a given outcome actually occurs, and the consequences to the program if that event, in fact, occurs. For the MDSE, Risk Management is the act or practice of dealing with these risk events, and includes:

- Assessing risk areas to the program: Assess risks that would significantly impact cost, schedule, or technical performance of the system should they occur, plan to mitigate their chance of occurring or their consequence should they occur, and manage the risk to a successful conclusion.
- Planning for the risk events: Provide a standard process to be used at each level of the SoS structure to ensure consistent and thorough identification, assessment, mitigation, management, and reporting of program risks.
- Developing risk-handling options: Focus on risk management philosophies that ensure an appropriate balance between risk management and a favourable impact on system life cycle cost.
- Monitoring the program to determine how risk event likelihood has changed: This includes quantifying the risk events at given intervals to evaluate their likelihood and/or program consequences.
- Monitoring the program for new potential risk events: This should be done regularly against both the high risk areas that have currently been identified, and looking for new risk areas that might arise as the system development evolves.
- Documenting the overall Risk Management strategy and risk event handling: Documentation is important to capture lessons learned for the current project as well as information for future projects.

The overall Risk Management process is illustrated in Fig. 1.14.

1.5 Overview of the Book

The overall purpose of this textbook is to introduce the student to Systems Engineering, and, in particular, Multidisciplinary Systems Engineering. We have arranged the book to begin with basic Systems Engineering principles and build up the overall practices of Multidisciplinary Systems Engineering throughout the book with the use of design problems, research problems, case studies, and discussion. The progression of the book is described below:

1.5.1 Chapter 2: Multidisciplinary Systems Engineering

Chapter 2 describes the overall need for a modern multidisciplinary systems engineering approach, which includes the introduction of System of Systems Engineering, the role of Software Engineering as part of the overall System Engineering process, as well as the need for disciplines like Test, Continuous Integration, Mission Assurance, and System Security. In Chap. 2 we introduce the system design problem that the students will incrementally design and will flow throughout the book, building up to a complete system design at the end. In Chap. 2 we introduce the first case study, a study of multidisciplinary techniques applied to experimental Biology, one of many case studies provided in Chaps. 2–10 on a variety of Systems Engineering issues that will be used throughout the book to help guide discussion and help the student understand the pitfalls of incomplete, incomprehensive, and/or just bad engineering practices.

1.5.2 Chapter 3: Multidisciplinary Systems Engineering Roles

Multidisciplinary Systems Engineers take on many roles throughout the design, implementation, test, deployment, and delivery of an engineering system. The purpose of Chap. 3 is to describe these various roles, which include Architect, Designer, Integrator, and various other roles as the project progresses. It is important to understand these roles and how they meet the needs of each stage of a system, for failure to provide these roles can be catastrophic to the overall success of delivery of a viable, usable, and operational system.

1.5.3 Chapter 4: Systems Engineering Tools and Practices

Modern Systems Engineering requires a host of productivity tools that allow the design, integration, and correlation of each of the architectures described above, and can be utilized by Software Engineers to define and guide software development throughout the system development. There are many standards utilized by commercial companies, as well as the Department of Defense. Each has advantages and disadvantages and the systems engineer may have a choice of which tools to utilize, or the methodology and tools may be dictated by the customer or management based on a variety of reasons, some technical, some financial. Chapter 4 will provide information on a variety of Systems Engineering productivity tools and methodologies to provide the student an adequate overview of what is available.

1.5.4 Chapter 5: The Overall Systems Engineering Design

At its heart, Multidisciplinary Systems Engineering is an interdisciplinary engineering process created to facilitate the realization of successful designs. It focusses on defining the requirements and required system functionality, utilizes these for design synthesis and system validation. Systems engineering is a structured development methodology that encompasses the entire system development process from concept development to production to operations. Chapter 5 is devoted to describing and providing insight into the overall Multidisciplinary Systems Engineering Design and describes how Systems Engineering considers business/mission needs, technical needs, performance needs, and all aspects/needs/goals of the system to provide the project customer with a quality product that meets all of the user needs.

1.5.5 Chapter 6: System of Systems Architecture Design

The study of System of Systems (SoS) Architectures is a discipline that is focused on the architecture and design of engineering systems; analyzing them in a holistic and integrated view of their strategic direction, business practices, information flows and technology resources. By analyzing and visualizing current and future versions of this integrated view, the SoS Architectures provides for the transition from current to future operating methods. This transition includes the identification of new goals, activities and all types of capital and human resources (including Information Technology) that will be required to improve financial and mission performance (strategic vision). In this discussion you will explore how to visualize and analyze SoS Architectures and how different views into the architecture allow you to ensure that all requirements are captured within the architecture design. The purpose of this chapter is to help you understand the principles and practices of architecture analysis and visualization. This includes discussions that describe Requirements Engineering, Architecture Development, and various methodologies utilized throughout the Systems Architecture process (e.g., Use Cases, Sequence Diagrams, etc.) [24].

1.5.6 Chapter 7: Systems Engineering Tasks and Products

Throughout the life cycle of system design, development, testing, and deployment, Systems Engineers perform many tasks and take on many roles. Each of these roles and tasks produces a variety of Engineering Products that are delivered to various "customers" throughout the life of the project. This chapter will provide descriptions and insights into the major task categories and the engineering products developed during each of these tasks and systems engineering roles. These include such tasks as Technical Planning, Business/Mission Analysis, Technical Coordination, and a host of others.

1.5.7 Chapter 8: The System of Systems Engineering Process

SoS Engineering provides the processes and oversight for the three main critical aspects of all programs: System Development, Site Installation and Checkout, and System Transition to Operations. The Systems Engineering process depicted is one of the primary influences on System Engineering and on the development of the System Engineering plans required for programs. Chapter 8 describes the overall SoS Engineering processes that are utilized throughout the development/operations life cycle in detail, providing information of the methodologies, principles, and practices required within a program structure are fully aligned to provide a full understanding the SoS Engineering technical development approach. At a high level, the programmatic system engineering process involves development of various program plans that will be used to direct and manage the design, implementation, and test of the program.

1.5.8 Chapter 9: Plan Development Timelines

While Chap. 9 provides a description of the SoS Engineering plan development processes, describing the inputs, outputs, and dependencies of various disciplines on plan development. However, it does not provide information on the phasing or timing of when each plan is required across the system development process. Chapter 9 describes the layout of each major project milestone and illustrates which plans must be in “initial release” maturity and “baseline” maturity during each major phase of program development. The point of Chap. 9 is to provide the student with the insight that plans are not made in a vacuum and require cooperation among all engineering and management disciplines to ensure that the plans are complete and meet the program needs across the entire program lifecycle.

1.5.9 Chapter 10: Putting It All Together: Systems of Systems Multidisciplinary Engineering

Here in Chap. 10 we put all the pieces together in the book and describe how to take a real, high-level, multidisciplinary, enterprise view of Systems Engineering. This includes discussions of top-down vs. bottoms-up engineering, as well as case studies that illustrate the results of not taking the multidisciplinary approach to SoS Engineering.

1.5.10 Chapter 11: Conclusions and Discussion

Chapter 11 is devoted to a providing the student with a final look at what happens when Systems Engineering goes right and what happens when Systems Engineering goes wrong. In addition, we include topics to provide the student with insight to current and future trends in Systems Engineering, like Agile Systems Engineering, an evolving field whose goal is to keep Systems Engineering in line with Agile and Extreme programming methodologies. Finally, we discuss the basic philosophy of how to make Systems Engineering useful. For at its core, the role of systems engineering is to make systems successful, not to be an end unto itself.

Chapter 2
Multidisciplinary Systems Engineering

2.1 Multidisciplinary Engineering for Modern Systems

Multidisciplinary Engineering will be a requirement for designers of modern systems in the near future. Much of the undergraduate Systems Engineering education is not effective in preparing engineers for the multidisciplinary approaches that will be required for system-of-systems integration and system optimization in the future. Commercial companies are becoming increasingly aware of the need for systems engineers, particularly systems engineers that understand a host of disciplines required to design, implement, and manage complex Information Technology systems. Engineering students must learn and adopt a breadth of disciplines to be ready for the systems engineering challenges of the future. Not only will Systems Engineering skills be required, but a host of other skills like Business Intelligence, Human Factors, Technology Integration, along with a working knowledge of Science, Technology, Engineering, and Math (STEM) skills. Systems Engineers can no longer become a stovepipe of systems engineering knowledge, and assume someone else will handle making sure their designs conform to the needs of other disciplines. Figure 2.1 below illustrates the confluence of these skills into the field of Multidisciplinary Systems Engineering [25].

> ***Question to think about:***
> *What actually happens when Systems Engineering does not provide a robust system solution?*

The study of Multidisciplinary Systems Engineering begins the journey of understanding the difference between studying engineering and becoming a practicing engineer. The purpose of this introductory Multidisciplinary Systems Engineering textbook is to help the engineering student begin this journey.

J.A. Crowder et al., *Multidisciplinary Systems Engineering*,
DOI 10.1007/978-3-319-22398-8_2

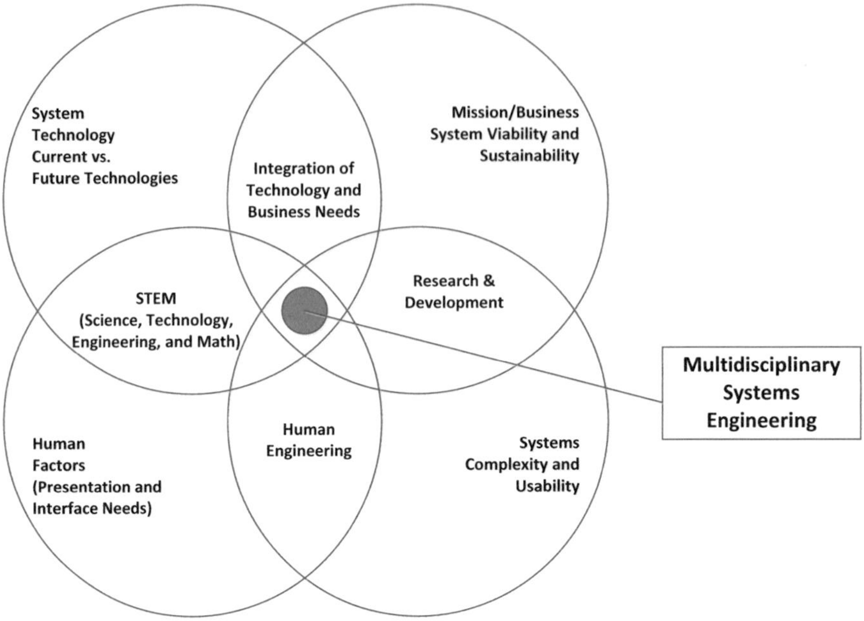

Fig. 2.1 Multidisciplinary systems engineering as a convergence of multiple skills

2.1.1 Multidisciplinary Approaches to Knowledge

Research shows that new knowledge is accomplished via natural human means and observations. For example: mental thought about observation or insights, the scientific inquiry process, physical human sensing of their surroundings, learning from results of actions, and experiences, while context is information, which adds to the characterization of knowledge, using terms and sentences, and gives it additional detail to focus the meaning. This knowledge is generally acquired via established forms of scientific research requiring the focused development of an established set of domain specific criteria, uniform approaches, detailed designs, and domain appropriate analysis, as inputs into what can be the potential solutions.

Multidisciplinary Engineering research literature clearly argues for development of strategies that can transcend and bridge the knowledge of any one given engineering discipline to another and engineers to enhance research collaboration between disciplines. Increasingly, cross and multi-domain research is more commonplace, made possible through a wide-range of available web based search engines accessible from ubiquitous numbers of devices, information content, and digital toolkits all part of the emerging Internet of Things (IoT) [26]. This environment creates the condition where large amounts of inadvertent cross-domain information content are exposed to wider audiences. Researchers expecting specific results to specifically focused queries usually end up acquiring somewhat ambiguous additional results and hence additional responses much broader in scope than was originally planned by the instigating query. Consequently, a resulting lengthy iterative learning process ensues along with numerous attempts at query refinement, until the truly sought after knowledge happens to be discovered. As a multi-disciplinary example Biomedical and Health care systems [27] generally access vast stores of medical research and clinical information content in attempts to gather detailed information and research about a particular topic or disease or condition.

Often digital and non-digital book searches can yield thousands of possible sources, most of which are not relevant to the context that a medical professional or non-medically inclined patient is seeking. Queries across many types of symptoms can lead to potentially hours of work and contemplation before an answer can be reached. This recursive-like refinement of knowledge and context occurs as user cognitive system interaction, over a period in time, where the granularity of information content results are analyzed, followed by the formation of relationships and related dependencies. These dependencies exist, even for a medical professional cross a number of domain boundaries. Even a doctor who has learned a great deal at medical school does not commit all medical content to memory and has to reach continually across discipline boundaries to achieve appropriate and many times life changing information content. Hence, underlying the ability to perform true data fusion within a domain is a significant challenge to create actionable knowledge from a continually expanding environment of vast, exponentially growing structured and unstructured sources of complex cross-domain data.

Therefore, Systems Engineering, as well as all engineering domain(s) requires new multidisciplinary methodologies to deal with the challenges of cross-domain content to more efficiently and more effectively discover and assimilate the most appropriate content for creating designs and architecture for systems that are viable enough to continue to advance research and compete in future complex environments in need of minimizing ambiguity and maximizing clarity.

2.1.2 Multidisciplinary Engineering as on Overriding Guide for the Design

As we discussed in Chap. 1, current and future system designs require System Engineering to embrace and utilize a multidisciplinary approach. The balance between depth of Systems Engineering principles and practices and a breadth of knowledge of the multiple disciplines (refer to Fig. 1.5) required to adequately design modern, complex systems must be evaluated throughout the system development lifecycle. Figure 2.2 illustrates this balance.

As expected, the overall goal of Multidisciplinary Systems Engineering is the successful design, implementation, delivery, and operation of a given engineering system. Project Management texts will frame success in terms of timescales, budget profiles, and resource utilization. And while not unimportant, these criteria do not reflect the delivery of a successful, viable, usable system [28]. Here, we couch success in terms of different entities: the Chief Multidimensional Systems Engineer, the Design and Development Team, the Multidimensional Systems Engineering Process, the overall design, and the end product (system). Figure 2.3 below illustrates the interactions between each of these entities.

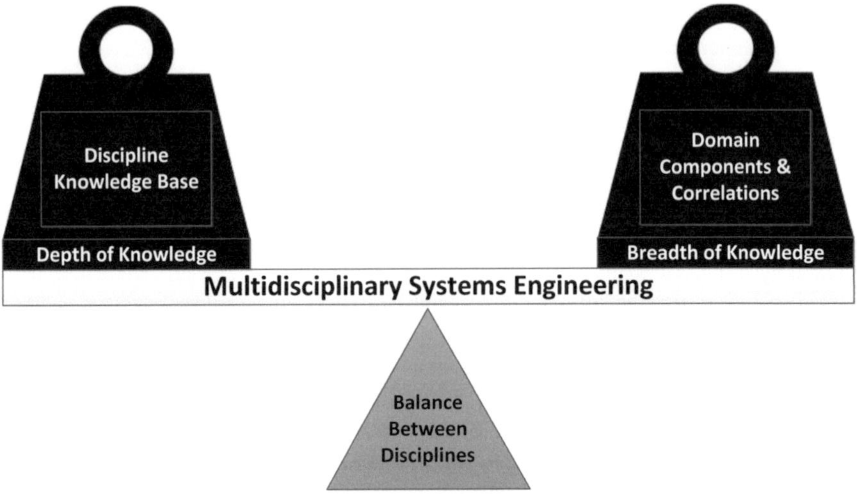

Fig. 2.2 Balance between systems engineering knowledge depth and a multidisciplinary knowledge breadth

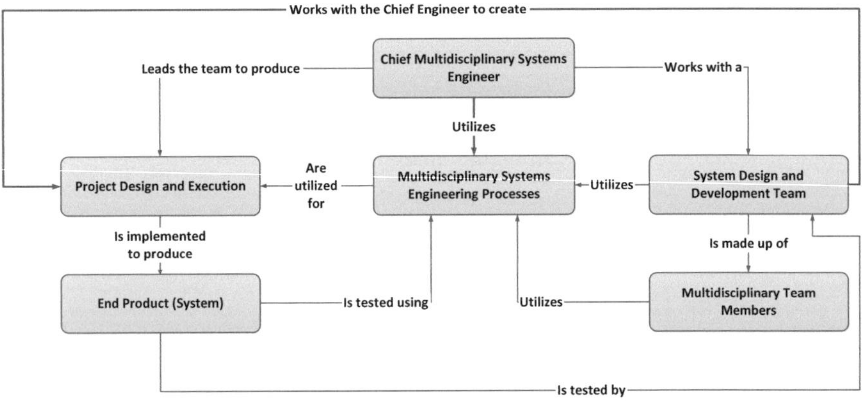

Fig. 2.3 The multidisciplinary success process dynamics

For such a multidisciplinary design and implementation process, success depends on the behavior of the development team and the overall ability of the team members to work together in the creation of the various system architectures (e.g., system, information, data, software, etc.), as well as the inherent way in which the design team employs a recursive mentality in their design process. For the Multidisciplinary Systems Engineer, project success is seen in terms of:

- An overall design process, not just an end product.
- The interrelationship between each of the steps in the design and implementation process.

- The balance between each of the engineering disciplines, making sure that all aspects of the design are equally considered throughout the system lifecycle (e.g., Human Engineering, Security Engineering, Systems Engineering, etc.).
- The overall level of team work and engineering skills, including communication.

These are inherently important to measuring the overall success of a project, since the measure of these factors is useful in determining the effectiveness of the team and whether this team makeup should be kept for other projects. The result is that success in Multidisciplinary Systems Engineering does not just depend on good designs and architectures, buy also on the socialization of team members across each project/design and the interactions between teams and team members to communicate lessons learned.

2.1.3 Multidisciplinary System of Systems Engineering

System of Systems Engineering began within DoD-related system development; however commercial companies are embracing the concept as they try to bridge legacy systems across companies and across geographically disperse stand-alone systems [29]. This may be through company acquisition, or across intracompany divisions. This drives the high-level definition of System of Systems characteristics: operational independence of System of Systems elements, evolutionary behavior as the systems are interconnected, and emergent behavior as coherent information flows drive the System of Systems architecture to become more than the sum of the system of systems elements [30]. The operational independence characteristic describes the boundary behavior of the connected System of Systems elements, while the other characteristics define the overall behavior, once the system is connected and operational. The last characteristic, emergent behavior develops due to complex interconnectivity of System of Systems elements that do not manifest themselves when each System of Systems element is operated in isolation. Understanding the dynamics of emergent behavior presents one of the primary challenges that Multidisciplinary Systems Engineering must face when designing and architecting System of Systems. The Multidisciplinary Systems Engineer must learn to understand the mechanics of evolutionary and emergent behavior, develop the metrics necessary to detect them, and establish methodologies for managing such behaviors. To that end, there are many types of System of Systems that the Multidisciplinary Systems Engineer may be involved in designing and implementing: the Directed System of Systems, the Collaborative System of Systems, the Acknowledged System of Systems, and the Virtual System of Systems.

2.1.3.1 The Directed System of Systems

Directed System of Systems are implemented for very specific purposes. It is usually build for long-term operations (e.g., satellite ground control system). The Directed System of Systems is intended to fulfill its specific purpose, but is expected to be expandable to encompass other purposes as the mission/business needs arise and are dictated in the form of changing requirements. Here, the Multidisciplinary Systems Engineer must design each element and bring the elements together to form a complete System of Systems. The elements of the System of Systems cooperate and are dedicated to the specific purposes specified by the overall customer.

2.1.3.2 The Collaborative System of Systems

Within the Collaborative System of Systems, system elements work together (or collaborate) to fulfill specific purposes that may be different from any of the individual elements of the System of Systems. Each element of the System of Systems decides separately how they will provide or deny service, allowing each element independent control over enforcing and maintaining standards. An example of the Cooperative System of Systems would the World Wide Web. In this System of Systems, the Multidimensional Systems Engineer from each element must collaborate and cooperate to create the System of Systems behavior required to fulfill its goals and objectives. There may be a formal agreement or an unwritten agreement such that it benefits all parties to keep the System of Systems goals and objectives fulfilled long-term.

2.1.3.3 The Acknowledged System of Systems

For the Acknowledged System of Systems, there are high-level objectives that must be fulfilled through the collective use of the Elements of the System of Systems and these must be agreed to and maintained by the collective System of Systems. In this model, however, each element remains independent, with separate ownership of each element, and each element retains its own objectives and purposes, operations, maintenance, etc. Any changes required to repurpose the entire Collaborative System of Systems must be based on collaboration between the over-arching System of Systems and each of the elements. Here, the Multidisciplinary Systems Engineers for each element carry a dual role, where each element has the responsibility to maintain its own goals and objectives that it must fulfill, while at the same time ensuring that the larger System of Systems also meets its goals and objectives. Here the Multidisciplinary Systems Engineer must manage a global system optimization, while continuing to manage their element-level system optimization.

2.1.3.4 The Virtual System of Systems

In the Virtual model of System of Systems, each element maintains complete control and there is no overall central management structure. The System of Systems required emergent behavior is dependent on relative mechanisms that the Multidisciplinary Systems Engineers together maintain, even if behind the scenes.

The emergent behavior may or may not be what is required to fulfill the requirement, needs, and objectives of the System of Systems customer. In many cases, they may wait to observe the emergent behavior that ensues, and then modify their objectives to match the observed emergent trends in the System of Systems. The Multidisciplinary Systems Engineer is much more concerned with their own element-level system, while behind the scenes trying to manipulate their elements to drive the overall emergent behavior of the larger System of Systems. This model is a very difficult System of Systems to facilitate.

2.1.4 Establishing an Effective Frame of Reference: The Heart of Multidisciplinary Engineering

Understanding the overall operational system environment is essential to understanding how to decompose the system requirements, needs, and goals into a viable System of Systems architecture. Based on the environment, the requirements, the performance needs, the external interfaces, the concepts of operations, constraints, and other applicable design considerations, a basic set of reference frames should be developed that guide the architecture design, system development, transition and operations of the System of Systems [2].

2.1.4.1 Analysis Frame of Reference

2.1.4.1.1 Logical Analysis

During Logical Analysis, the required System of Systems functionality and overall system behavior is analyzed. This analysis defines, selects, and designs the logical architecture required to create a Logical Framework from which design can be assessed for compliance with all system requirements for all operational scenarios required and for long-term System of Systems operations. The Logical Analysis will include Trade-off analyses between requirements and proposed solutions. This Logical Analysis will include creation of Functional Models (functional decomposition) of the system requirements, Semantic Models that describe the relationships between information and data required within the system (e.g., Data Flow Diagrams, Entity-Relation Diagrams, etc.), and Dynamic Models that describe state transition diagrams and activity required to describe and meet the overall system requirements. The Logical Architecture that flows from this analysis is a set of technical concepts and principles that support the logical operations of the systems [31].

Question to ponder:
What is the difference between the logical architecture lay out of the systems vs. the physical instantiation of the architecture?

Improper handling of the Logical Analysis and subsequent Logical Architecture is the reason for many system design and implementation failures. Some of the common problems with improper Logical Analysis are:

- **Improper handling of Mission/Business needs**: the Logical Analysis should take into account the customers' operational concepts. If these are not well understood or folded into the overall Logical Architecture, the resulting developed system may not be useable for the purposes for which it was originally intended.
- **Improper handling of system requirements**: there are times Multidisciplinary Systems Engineers will pay attention to some requirements more than others because they are well versed in designs using some requirements and not well-versed in others. The consequences are that the ensuing Logical Architecture defines a system the Multidisciplinary Systems Engineer is comfortable with, not one that meets the requirements and needs of the customer and system operators.
- **Improper Functional Decomposition**: if the Multidisciplinary Systems Engineer creates of too many functions within a given level of the architecture, or too many levels of architecture, the number of internal interfaces and data/information flows increases, overly complicating the overall design, implementation, operations, and maintenance of the system.
- **Improper consideration of inputs/outputs**: Many systems architects consider only the functionality and subsequent actions needed to support the functionality while not considering the system inputs and outputs as part of the overall, integrated design, deciding, instead, to deal with the system inputs/outputs as an afterthought later. The Multidisciplinary Systems Engineer must realize that inputs and outputs are an integral part of the overall design and must be included at every step in the architectural design process.
- **Improper emphasis on static functional decomposition**: static decomposition of the system is important, but is a small part of the overall architectural design and should only be performed after the functional scenarios have been created and folded into the Logical Architecture. The static decomposition should be one of the last operations that the Multidisciplinary Systems Engineer deals with in the architecture design.
- **Improper handling of Governance and System Management**: Strategic Monitoring and Orchestration of the system (Governance) and Tactical Monitoring of the system (System Management) are often mixed and handled as one set of system functionality. Behavioral management as well as temporal management of the System of Systems are related but must be managed within the design individually. These are very distinct functions and mixing them will result in improper use of both within the overall Functional Design.

Question to think about:
What do you think the effect is on the usability of the system, when system monitoring is not included in the design?

2.1.4.1.2 Risk Analysis

Risk analysis is important to assess the potential technical risks associated with the System of Systems environments, technical constraints, proposed new technologies, or anything that could pose technical issues throughout the program lifecycle. It is impossible to assess 100 % of the potential risks to the system (e.g., predicting a category 5 hurricane hitting New Orleans). At the System of Systems level, any legacy system that will be used in the overall design must be evaluated for impacts to the optimization and overall operations of the System of Systems. For the Multidisciplinary Systems Engineer, risk analysis must involve Systems Engineers from each of the elements of the System of Systems, along with Subject Matter Experts (SMEs) from each of the technologies to be employed throughout the System of Systems in order to adequately assess the technical risks to the program. Both the likelihood of occurrence and overall system impacts posed by the use of the legacy system are important risks to be determined and understood and then decisions must be made as to how to handle the potential risks. More will be discussed in later Chapters and Risk Management is normally its own course within the overall Multidisciplinary Systems Engineering curriculum.

Question to think about:
What effects are there on the risk posture of a project when Inputs/Outputs are ignores?

2.1.4.1.3 Decision Analysis

Decision Analysis involves determining the methodologies and metrics that will be utilized throughout the System of Systems to assess performance against the customers' objectives. Some of the issues to be analyzed are:

- What metrics and measures of effectiveness (MOEs) should be collected?
- When are the appropriate times to take metrics?
- Which system nodes should be monitored throughout the System of Systems?
- What kind of root cause analyses should be utilized for problem identification/resolution?
- What types of Alerts, Warnings, and Event notification should be utilized throughout the System of Systems?
- What changes might impact the System of Systems, to include changes in technology as well as mission/business need changes?

2.1.4.1.4 Data/Information Analysis

The purpose of Data/Information Analysis (which is part of the discipline of Information Engineering) is to determine the data/information objects that are required to support the overall system-level requirements, objectives, goals, user scenarios, etc. Understanding the types, quantity, and quality of the data that are required by the system can radically change the overall design of the system. "*Big Data*" has become a major field of study in engineering over the last two decades, as the availability of analysis techniques increases, and the need for more detailed analysis grows (e.g., cybersecurity), more and more data are required to feed the analysis and decision engines of today and the future. If you are building a toaster, the quality of data required by sensors is trivial compared to, let's say, our GPS systems. In GPS, quality of data/information is paramount since the world relies on the accuracy of GPS data for everything from how get to the nearest Starbucks, to precision timing of transactions on Wall Street, to military operations; Starbucks being the most imperative one on the list, of course.

Not only does data volume needs to be considered, but the heterogeneity of the data must be considered, as well as who needs access to which types of data. In performing Data/Information Analysis, the Multidisciplinary Systems Engineer should be determining a host of data/information characteristics that must be taken into account by the Multidisciplinary Systems Engineer in the overall architectural design and implementation of the System of Systems. The System of Systems Data/information ontology should be developed with subsequent data/information taxonomies for each element of the System of Systems. Figure 2.4 is an example of an ontology created for multi-media data/information. Some of the considerations are:

- Types of data (textual, scientific, etc.)
- Data representations (e.g., structured, semi-structured, unstructured, etc.)
- Precision required for each type of data/information
- Overall data throughput for the System of Systems
- Message types and quantities required across the System of Systems information system

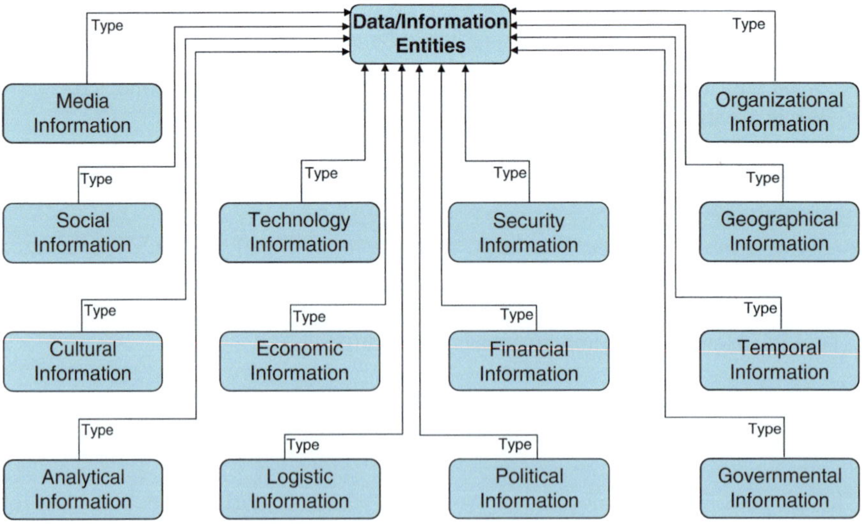

Fig. 2.4 Example of a data/information ontology

- Usage of each data type
- Data/Information interfaces required (both external and internal interfaces)
- Data/Information protocols that are needed/required for the System of Systems
- Data transformations required; possibly due of the use of legacy systems, or due to external interface requirements
- Data/information storage requirements
- Data/Information sources (where does all the data come from, both internally generated data/information as well as externally generated data/information)
- Data management and maintenance
- Data/Information presentation requirements

Question to think about:
What are the key differences between data ontology vs. storage structures for data?

2.1.4.2 Technical Frame of Reference

For System of Systems design, implementation, integration, and testing generally requires a distributed development, with each element of the architecture having a separate development team; while the Multidisciplinary Chief Systems Engineer must oversee the distributed efforts to ensure a smooth integration of the overall system. In order to accomplish this, a common technical frame of reference must be established among the various elements. The common technical frame of reference ensures the element-level technical solutions meet all requirements, goals, and mission/business needs in terms of technical performance, data/information needs, which then allows Integration, Verification and Validation, Transition, Operations, and Maintenance to be accomplished without major rework between elements. To that end, there are a number of technical frames of reference that must be established by the Multidisciplinary Systems Engineering organization for the System of Systems program.

Question to think about:
Without common technical frame(s) of reference, will the system actually integrate and operate without extensive cost and schedule impacts?

2.1.4.2.1 Requirements Decomposition and Allocation

The challenge with distributed development is ensuring that requirements are properly decomposed and derived for each element, that requirements are not missing or misunderstood, and there exists a clear roadmap that establishes how the element-level requirements and their subsequent decomposition can be used to demonstrate compliance with the System of Systems-level requirements.

Table 2.1 Requirements decomposition and derivation criteria

Characteristic	Description
Requirement prioritization	Comparison weighting based on overall mission/business needs
Requirement source	Where did the requirement come from: customer requirements, concept of operations, constrants, laws, and/or agency restrictions (e.g., FAA)
Requirement usage	Technical functionality alternatives that can be used to execute this requirements
Requirement impact	Performance consequences, potential risks the requirements pose on the system, architecture impacts
Requirement user scenario	How will this requirement be utilized by the operational system and how the people who will operate the system will benefit by this requirement

Requirements decomposition and derivation is one of the most important aspects of system design, where bad requirements decomposition and derivation is almost impossible to overcome during the development phase without a major amount of rework, including redesign of the various architectures and new development efforts. Bad requirements will have major impacts on the overall cost and schedule of the project. Creating a common requirements framework and management scheme for all requirements decomposition and derivation among each element, therefore is paramount for program success. Table 2.1 illustrates an example template for all requirements work. There are many electronic tools to manage requirements; the list below shows a few:

- IBM—Rational DOORS Next Generation®
- Intland Software GmbH—codeBeamer Requirements Management®
- Hewlett-Packard—HP Agile Manager®
- Polarion Software—Polarion Requirements (2014 SR2)®
- TechExcel—DevSpec®

Question to think about:
Are requirements necessary for a system development?

2.1.4.2.2 Data/Information Decomposition and Allocation

In order to understand how the data/information in the system should be allocated and utilized across the System of Systems, as well as the overall capabilities required by the data/information at each level in the architecture, the Multidisciplinary Systems Engineer must understand the following:

- **Data/Information Quality**: here we define quality in terms of the accuracy, compliance, and completeness of data/information objects and their overall ability to satisfy the system requirements, goals, objectives, etc. This includes topics like data usage, data rate of change within the systems, and synchronous vs. asynchronous data issues.

- **Data/Information Structure**: How will the data/information be organized across the System of System architecture, including the structure to be utilized at each level?
- **Data/Information Architecture**: This includes information structures, processes, and organizations (who is responsible), as well as storage alternatives for data at the System of Systems Enterprise level.
- **Data/Information Governance**: The roles, responsibilities, policies, procedures, and governing laws that are required to manage all data/information across the System of Systems.
- **Data/Information Security**: This involves identifying the security requirements on the data in the system, and who has access to which data/information in the system, and how to manage authentication and authorization for user/process access to protected information.

2.1.4.3 Technical Management Frame of Reference

The role of Technical Management within the System of Systems development is to understand the technology base for the program. For long-term programs, Technology Management is a continuous set of processes and strategies that the Multidisciplinary Systems Engineer utilizes to manage the use of technology during development, with a goal of optimizing the overall system technology to meet the Quality, Cost Profile, and Schedule for the program. Technical Management utilizes many concepts within its overall frame of reference:

- **System of Systems Technology Roadmap**: what technologies are available when, and how will they be introduced into the system over the life of the program.
- **System of Systems Technology Strategy**: the role technology plays in the overall success of the System of Systems program.

2.1.4.3.1 Planning Management

Technical Planning Management involves organizing, controlling, and executing the processes, procedures, protocols, and activities required meet the requirements, goals, needs, and performance of the System of Systems. For the Multidisciplinary Systems Engineer, technical planning requires constant attention from the early beginnings of a program, to its ultimate retirement when a new system is built and transitioned. This includes optimizing the allocation of the necessary inputs, functions, and outputs to integrate them to meet all pre-defined System of System objectives. Planning Management has been around for thousands of years and was probably put in place by early Egyptian Pyramid builders. Figure 2.5 below is a picture of a carving discovered from Rome in 113 AD, depicting roman soldiers building a fortress.

Fig. 2.5 Picture of carving showing a roman building project from 113 AD

2.1.4.3.2 Requirements Management

As we have discussed several times and will continue to discuss, because it is ***that*** important, the handling of requirements is one of the single-most important activities the Multidisciplinary Systems Engineer must clearly understand and handle well if program success is at all a priority (of course, it ***always*** is). Managing the Technical Requirements is necessary to define the technical issues that the Multidisciplinary Systems Engineer must deal with in creating the overall architectures and designs required for successful delivery of a System of Systems that meets all requirements, goals, and needs. Part of Requirements Management is to make sure the requirements at every level in the systems are:

- **Defined Precisely**: Technical Requirements (all requirements really) should be clearly worded, sufficiently detailed to eliminate ambiguity, necessary (related to the mission/business needs), and testable in relation to the tier of the system.
- **Analyzed for Feasibility**: Part of Requirements Planning is to determine the feasibility of the requirements at each level of the System of Systems, ensuring that there are no conflicts in the requirements that will make execution of the system problematic. This includes an impact analysis of the major requirements and their impacts on the overall architecture.

2.1.4.3.3 Risk Management

Once program risks have been determined and categorized, as discussed earlier, program risks must be managed throughout the lifecycle of the system to ensure system success in the event any of the identified risks are realized. Risk Management involves determining the processes, methodologies, metrics, tools and Risk

Management infrastructure which is appropriate for each System of Systems design and implementation. Having a standard methodology for Risk Management across each element in the System of Systems design allows seamless communication of mitigation strategies for each program risk. There are many Risk Management COTS S/W that is available, including freeware packages. Several are listed below:

- Metric Stream—Risk Management Software Solutions®
- SAI GLOBAL—Enterprise Risk Management (ERM)®
- Integrated GRC—Acuity Risk Management®
- SwordActiveRisk—Active Risk Manager®

2.1.4.3.4 Data/Information Management

Data/Information Management runs through the entire system lifecycle, from initial requirements decomposition/derivation/allocation through system retirement (cradle-to-grave). Whether all data is contained within relational databases, contained as individual data items, or as part of flat files, all data must be adequately managed throughout the System of Systems in order to track and control data changes as the design and or system environment changes.

Data and information management must the integrated into the Configuration and Change Management processes used by Multidisciplinary Systems Engineers. This ensures that all of the data/information, archive information, database schemas, and data/information flows throughout each architecture level are controlled and managed during the development cycles. This management and control ensures that a central repository is available to all elements in the System of Systems development. This includes the date of change for the data formats on programs with multiple development periods. Note: This can be mitigated through some of the modern data definition schemas (e.g. xml), as long as the usage rules are properly defined.

The Multidisciplinary Systems Engineer creates methodologies and constructs to create and manage Metadata,[1] which is utilized to provide identification and organization of System of Systems data/information. Formal Data/Information Management allows enhanced collaboration across System of Systems elements by identifying uses and formats required by each element and storing data in one place, transforming it to its intended format as required by the element.

2.1.4.3.5 Configuration and Change Management

Configuration and Change Management is involved in managing changes to the System of Systems. This includes:

- Managing change requests
- Planning changes—application of effective change dates
- Implementing
- Assessing change impacts

[1] Metadata is information used to identify and manage data/information. Metadata is used to organize data/information and resources across the system.

The main goals of Configuration and Change Management provide traceability of changes across the System of Systems.

Questions to think about:
Even with Change Management utilized within a system development, does the technical frame of reference ensure that Change Management results are valid?

Even with Change Management utilized within a system development, does its existence ensure that the technical frame of reference will be adhered to?

What must happen to ensure Configuration and Change Management are effective from an MDSE point of view?

2.1.4.3.6 Interface Management

Management of all internal and external interfaces is critical for the Multidisciplinary Systems Engineer. Understanding all of the interface interactions, properties, protocols, and data/information formats is necessary not only for architectural design, but also is essential for integration of the System of Systems elements and testing of the System of Systems. The challenge for the Multidisciplinary Systems Engineer is specifying and designing interfaces in a way that can be utilized by all elements in the System of Systems, while adhering to industry standards. Interface Management is utilized to track all System of System boundaries and how they interact with each other (both external and internal). Interface interaction descriptions describe the operational entities used by services during all System of Systems operations.

Question to think about:
What happens in system development when interface management and requirements are not defined in terms of Systems Engineering, but in terms of Software Engineering?

2.1.5 Creating a Unifying Lexicon

Ensuring term definitions are unified across the entire System of Systems is essential to ensure that there is no confusion as data flows between elements and within element of the System of Systems [32]. The Multidisciplinary Systems Engineer needs to establish a team early on to establish a unifying term and data lexicon before architecture design has begun. When terms and definitions are not properly defined, risk to the development increases as well as potential system failure(s). A good example from history occurred with the Mars Probe in 1999.

One team was using Metric units for their element, while another team (on another element of the System of Systems) was using English Units. Unfortunately data passed between these elements for a critical operation on the Mars probe, which led to a catastrophic failure during the landing operation (a tragic mistake since *AAA* doesn't go out that far).

2.2 Software's Role in Multidisciplinary Systems Engineering

Software plays in integral part in Multidisciplinary Systems Engineering. The Software Systems Analyst role is necessary to determine whether the System of Systems architecture is implementable in software such that all the System of Systems requirements can be met. This includes analysis of hardware required to host the software, memory usage, external and internal interface traffic, and data/information flows throughout the System of Systems. This includes development of the Service Architecture and the Software Fault Architecture that will be required to assess the Reliability, Maintainability, and Availability (RMA) for the delivered System of Systems. The Software Analyst Engineering role within Multidisciplinary Systems Engineering is a very distinct role, which is to elicit, prioritize, suggest, and negotiate the decomposed/derived/allocated requirements in order to simplify and optimize the envisioned software design as much as possible. The Multidisciplinary Systems Engineer must be able to think like a Software Systems Analyst in order to create a System of Systems Architecture that not only can be realized in software, but also seamlessly integrates across the System of Systems elements [15]. The Software Systems Analyst role within Multidisciplinary Systems Engineering facilitates collaboration between systems and software engineers and allows creation of overriding principles for integration, interoperability, modifiability, and expandability of the System of Systems final product. Figure 2.6 below illustrates the Systems vs. Software roles as well as the combined roles the Multidisciplinary Systems Engineer must embrace [33].

2.2.1 Computational Thought and Application

Multidisciplinary Systems Engineering developed out of the need for foundational science in engineering across disciplines/domains, and invariably the systems developed and employed specifically for those domains. Understand that many disciplines reach out and touch other disciplines out of a desired need. The desire to automate biological and chemical science required software. The desire for automated mechanical systems in the auto and oil industries required software. Over time, through observation and implementation, the software engineering discipline instituted pattern development and recognition techniques for common and optimize development and implementation. Software engineers today, with longevity,

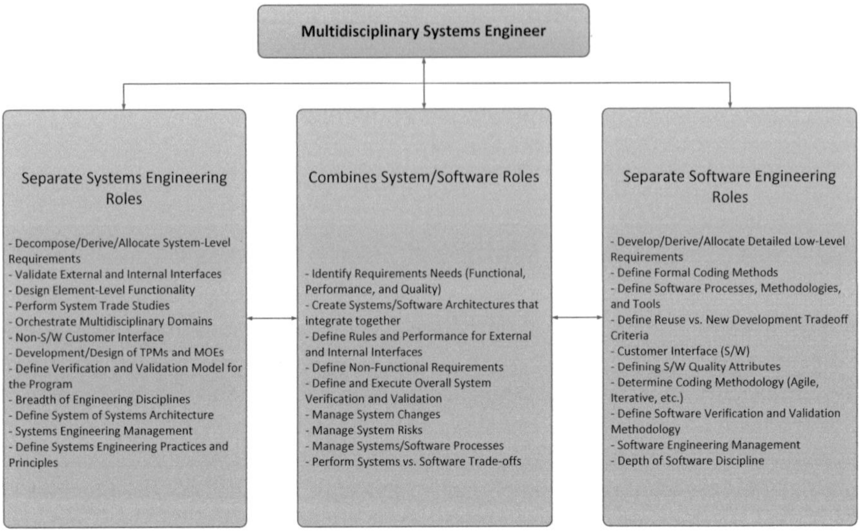

Fig. 2.6 Systems vs. software roles of the multidisciplinary systems engineer

have an uncanny ability honed over years to rapidly find the common denominators needed in a design and provide a design solution. The thought processes and mechanisms together are known as Computational Thinking (CT). Simultaneously, over the last few decades, via the Moore's Law expansion, many other disciplines developed their own improvements from the inside-out or Interdisciplinary, as seen in the Chap. 1, Fig. 1.4, as opposed to common improvements across disciplines (e.g. Multi- Trans-). This emphasizes the difference: Multidisciplinary is the act of applying methods from two or more disciplines to single topic within one discipline, whereas, Transdisciplinary, a higher form of Multidisciplinary, focuses on the science and engineering required to engineer for commonality across two or more disciplines. This fine-grained nuance is essential to the development of the right frame of reference to solve difficult problems and develop complex systems (System of Systems) within a discipline you might not initially fully comprehend.

2.2.1.1 Recursion's Role in Layering Abstraction

One of the difficult concepts for software engineers to initially understand is the time where they are exposed to the difficult concepts required for effective recursive engineering. Although not a difficult topic to understand, recursion can be a struggle to implement effectively. You will find many definitions of recursion. Recursion simply put, is an iterative process. The end state of a recursive process is an optimal solution based upon a set of thresholds defined as part of the criteria for starting the process in the first place. The inherent artistic nature of a good recursive process is one that most effectively processes all of the items and parameters while the process continues to unfold. Recursion is well-known in many disciplines where repetition is required. Repeating the same method over and over without the need to continuously understand the output of each of the iterations is what we would liken to the simplest form of recursion. We will discuss what is at the core of achieving successful optimization of any Multidisciplinary Systems Engineering design, a concept called Computational Thinking (CT).

2.2.1.2 Application of Computational Analysis and Thinking

The successful and most optimized applications of recursion, which includes iteratively optimized processes through the use of *abstraction*, requires determining the relationships between the different layers as the process recursively iterates through them, while simultaneously adjusting the parameters of optimization at each iterative layer. Abstraction is at the core of what is known as Computational Thinking, which is an inherently humanistic inquisitive process like 5WH, a term used by many for Who? What? When? Where? Why? and How? To achieve effective CT we apply 5WH and research in Cognitive Psychology to help you understand and apply Computational Thinking processes to be effective during engineering multidisciplinary systems on any project. As we have discussed, engineering across disciplines requires one to take into account many parameters and layers of abstraction in order to successfully achieve a solution. This cognitive process is known as Recombinant kNowledge Assimilation (RNA), comprising a set of instructions underlying learning algorithms [17], each using 5WH: discovery, decomposition and reduction, compare and contrast, association and normalization. Figure 2.7 depicts the high-level process occurring as one evolves through the set of cognitive RNA instructions analyzing, iterating and recombining abstraction layer content.

Discovery sub-process obtains content from the information domain; the content is decomposed and reduced to a threshold of understanding within the temporary knowledge domain. Subsequently, it is compared and contrasted to additional iteratively discovered content. Subsequently, relationships of understanding are made via association combined with normalization which finalizes the understanding of the new abstracted layer within the Knowledge domain. Effective recursive and

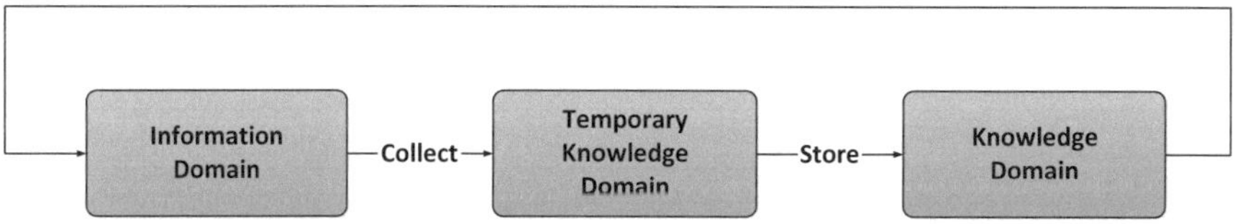

Fig. 2.7 Knowledge recombination domains

iterative optimization ends up as a feedback loop, with new Information Domain content utilized for the next iterations. This emphasizes the most important question a Multidisciplinary Systems Engineer working in any domain must ask, "How do I know when I have achieved an optimal level of abstraction for my system?", "How do I know when I am done?"

Questions to think about:
What happens when Systems Engineering is not complete relative to:

(a) *The system design?*
(b) *Hardware and software development?*
(c) *Cost and schedule?*
(d) *Risk?*

2.2.1.3 System Optimization Via CT Automation

The previous paragraphs introduced you to the underlying constructs and processes for using Computational Thinking. However, to answer the questions posed in the previous paragraph in terms of what is occurring during System Design optimization we will focus on the last RNA instruction, Normalization. As each layer of abstraction is defined and the relationships between the parameters of each of the iterations are understood a certain simplification or generalization naturally occurs. The evolving generalization improves after each of the iterations and is evolving into a common way of approaching the process until that final iteration matches the design and engineering thresholds one is attempting to attain. In a physical system design instantiation this might include a set of tolerances or design range in which a machine must operate within. The CT infused systems engineering design actions you are performing are automating the layered abstractions into a common approach, when implemented, become the optimized common system processes. Therefore, whether optimizing System Engineering designs within a discipline or across multiple disciplines evolves to enhanced set of common automatable solutions mechanized by Computational Thinking.

2.3 The Psychology of Multidisciplinary Systems Engineering

Multidisciplinary Systems Engineering requires a paradigm shift in the classical Systems Engineer. As programs become more and more complex, and as System of Systems become more prevalent, we must embrace change (even though it seems change is hard for engineers in general). The psychology of Multidisciplinary Systems Engineering is all about a change in Systems Engineering which will address the Multidisciplinary aspects of modern systems.

2.3.1 Resistance to Change

If change has always been an integral part of life, why do we resist it so? Why, in every generation, do we have major resistance to change? *It seems obvious that there should be thought put into change theories as we ask for people and environments to change.*

Some key components to encourage change are empowerment and communication. People need time to think about expected change. As we discussed earlier, people are good at change in order to master or improve their world or environment. When people remain agile they are better at being agile. When change is part of the regular process then one becomes agile. Alternatively, when one is used to doing something exactly the same way or systematically then it becomes the way one likes to operate.

Why, in particular, do some engineers not like change? When asking this question it seems obvious to consider education. What is it that has been required of engineers in the past and present and what will be required of them in the future, particularly for Multidisciplinary Systems Engineers. Juan Lucena [34] hypothesizes connecting engineers' educational experiences with their response to organizational change and offers a curriculum proposal to help engineers prepare for changing work organizations. As our technology increases and our work world becomes more agile it makes sense also that soft skills will become more and more important for engineers. Multidisciplinary Systems Engineers must organize themselves to optimize performance with soft skills to embrace the multiple domains they must utilize.

Trust is the most important factor in change. Trust helps to balance fear which is often the root of most resistance. Who in their right mind will blindly make changes without trusting others? Agile methods increase trust by increasing transparency, accountability, communication and knowledge sharing [35]. Iteration/sprint planning methods give members visibility on requirements, individual assignment, and agreed estimates. People get the information at the same time. The daily stand up provides visibility and transparency so issues can be addressed immediately. People will know if someone is behind. The sprint/iteration retrospective provides transparency and visibility regarding goals. This agile design builds trust among the team member as they are able to see others trustworthiness and competencies as they continue working together.

Crawford et al. [36] suggest that employees who had higher creativity, worked on complex challenging jobs and had supportive non-controlling leadership, produced more creative work. If workers see that their ideas are encouraged and accepted they are more likely to be creative. Empowerment and trust encourage creativity which encourages change which encourages agility [35].

Think about the complexities that people bring to projects. Think about projects of many teams made up of many people, virtual teams, many sites, many locations, many levels of expertise and experience, many cultures and places that people live. Then add in technology, many technologies, changing technologies, developing and revolutionary technologies. It becomes more and more clear that soft skills are incredibly valuable to engineers and particularly to agile teams. Azim [37], shows that up to 75 % of project complexity has to do with the human factor or the people in the project. They claim that soft skills are important in the implementation of plans. Soft skills are clearly important in any complex project and in all phases of change previously discussed, particularly in the commitment and transition of change.

Soft skills include organizational, teamwork, communication, and other people based skills. As technology matures and new technologies emerge it is imperative that the teams and people in the teams become more agile. Not only is technology constantly changing so are people. It seems ever so important that the soft skills come with the hard skills. Multidisciplinary Systems Engineers must be willing to embrace significant change in the industry in order to be effective with large System of Systems designs.

Question to think about
What happens to the company that fails to address change?

2.4 Case Study: Application Multidisciplinary Systems Engineering (MDSE) for Information Systems Applications to Biology

This case study is used to show how Multidisciplinary System Engineering practices should be applied to solution sets for something other than a traditional information management system, weapon system, power system, etc. The tools needed to develop a system used by biologists are relatively the same as another other system. The need for top down analysis, development of requirements and architectural solution remain the same, and the application of MDSE will enhance the outcome for the biologist's system. The following case study outlines a premise/argument for the use of multi-, trans-disciplinary engineering techniques to implement and then upgrade a system solution in the field of Systems Biology. The following materials provide the background needed to assess if MDSE can be/should be applied for the development of a system used in biological engineering.

Biological Engineering has only recently emerged from an applied science to its current status as a legitimate field of engineering professional practice. Biological systems exhibit a number of properties that challenge our traditional ways of thinking. Among the most dramatic are time and age dependent changes in form, composition and behavior, including evolution, adaptation, competition, emergent properties and others. Many biological processes respond to external stimuli such as light, temperature, pressure, population density, nutrition, etc. in sometimes-unpredictable ways through self-regulation, aggression, and learned or genetically derived anticipatory actions. Rarely can we design one element of a process without deep connections to surrounding elements [38].

Today, there exists much discussion about the need for transdisciplinary application of today's engineering solutions [27], as well as, the lack of use of transdisciplinary engineering theories in enhancing and augmenting the field of Systems Biology [39].

Hence, the goal of this section is to explore multi-, trans-disciplinary options for Systems Biology, and use those options to provide an analysis and potential suggested solutions to a specific set of Systems Biology requirements and problem set.

In biology, inclusive systems and interactions tend to produce more robust and sustainable outcomes—in biology the operative word is AND versus the traditional design choice OR in engineering. In spite of the great advances in biology and biological engineering science, designers are faced with high levels of uncertainty and many unknowns in their design efforts. Chaotic behavior is "normal" in biology. Every day we are surprised by discovery of previously unknown interactions between organisms and their environment, and among organisms in a population. The biological world is constantly affected by environmental and ecological pressures within and beyond what ecologists call the "natural range of historic conditions." With all this chaos, uncertainty and unknowable stuff, there are fundamental principles that we can use as anchors to form trends to follow function in biological engineering design."

Systems Biology is wrought with high levels of uncertainty and many unknowns as well. Antonsson [40], stated that the advancement of design theory requires "investigation into aids for the design process." The need for disciplines like bioscience and design science to investigate beyond the boundaries of their existing design concepts to create aids in advancing the design processes is great. Nowhere is this more self-evident than within the bioscience discipline. Almost to the point of understatement, are the stimulating discussions that come from scientific methods and concepts currently under debate in bioscience, Evolution versus Intelligent Design (ID). Scientific methods when applied to design theory, as in creating data and design process abstraction, do not stimulate the kind of fervor generated by Evolution versus ID. However, they commonly stimulate discussion surrounding the transdisciplinary paradigm concepts regarding mutually sharing of discipline methods and subjects between them.

Hence, we introduce multidisciplinary concepts and apply them to the problem set to achieve a specific set of qualitative solutions. Engineering disciplines which are either more mature, or have generated solutions to similar problem sets within their domain, can potentially provide added efficiencies and benefits because of similarities to requirements, designs, and problems in other domains.

2.4.1 Discipline Background Investigation

Examining the use of specific theories, methods, and practices known as First Principles lays the foundation of understanding. This must be performed before delving deeper in processes and methodologies, as well as addressing potential solutions to specific discipline problem sets. Once a foundational layer of First Principles is laid a specific set of requirements can be undertaken. In the following example we will examine a specific architecture; processes, methodologies and mechanisms used in a system developed to perform comparative System Biology analysis and look for potential optimizations.

A common mechanism for performing System Biological study is the use of micro-cell arrays to analyze cell growth to help determine cause and effect. A standard way of performing oncology research without the use of human cells is the use of less complex and more readily available yeast cells. A yeast processing system has several key components: Cell Array Analysis, an Experiment Management System, and Lab Automation in the form of robotics. A cursory analysis of disciplines in use within this system yields: image analysis, software, and hardware, along with command and control systems for the robotics. Each of these disciplines now requires a threshold of study of first principles to obtain a reasonable understanding in order to determine an initial system-level Discovery and Decomposition.

As part of our increased transdisciplinary study of foundational content relative to this new unknown system we look to recognize/identify which disciplines could provide certain optimizations related to our system. Upon doing so, we find that our study time should include theory and applications from the fields of Photogrammetry for the potential optimization of the micro-array system, Software Engineering for the software architecture and potential specialized mathematics for the imagery and analysis. Next, we begin using the application and theories from within the disciplines: software and hardware Systems Architecture, analyzing current workflow or business logic processes in use, and perhaps examining and augmenting the current mathematical formulations and methodologies to improve the accuracy and efficiency of the current cell array system design [41].

Photogrammetry [42] is a remote sensing technology which incorporates methods from many disciplines, including optics and projective geometry, to determine the geometric properties of objects from photographic images. Upon detailing the software system it becomes apparent that the potential optimization benefits of applying Software Engineering methods and theories to this Systems Biology analysis system can derive from componentization theory and Object Oriented Design (OOD) [16], which has been on the forefront of the information age since the late 1980s. Additionally, it also well known that the use Service Oriented Architectures (SOA's) to administer functionality, organize data and normalize workflow or business processes within a system, can have potential benefits for this system. Both Object Oriented Design and Service Oriented Architectures being methodologies key to the Multidisciplinary Systems Engineering approaches to System of Systems designs.

2.4.2 Current System Design: Experimental Approaches

An example system diagram of a semi-automated yeast processing system is depicted in (Fig. 2.8).[2] From left to right the graphics below describe an example system design within a systems biology lab. The system uses a number of paradigms and approaches to as a goal to increase the ability to determine phenotype's

[2] Depicts a representation of processing environment from the University of Alabama, Birmingham Hartman Genetics Lab (circa 2007).

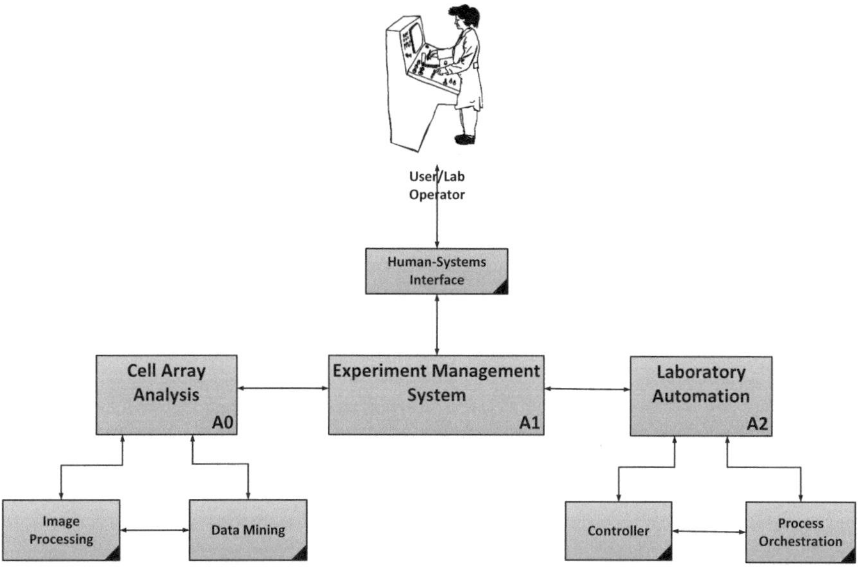

Fig. 2.8 Yeast processing system diagram

(specific outcomes) thru a systematic analysis approach of Haploid deletion strains of yeast (monitoring of cell growth), Cellular Array Technology Development, using Perturbation strategies (pair-wise approach to Chemical sensitivity and Genetic mutation combinations) and by Quantification of Interactions; determining the effect of gene deletion on phenotypic response to perturbation [43]. An analysis of the processing part of the system yields an example UML diagram depicting the state transitions of the current cell array yeast processing (Fig. 2.9).

2.4.3 Current Hardware and Infrastructure

The collection and processing, accuracy and measurement of the data are our focus with respect to the hardware. The hardware consists of a scanner, USB interface, PC hardware, and agar plates made up of Petri dish and nutrient based gelatin like solution. An expressions scanner is used for data collection. The data samples are scanned at 140 DPI for agar plates containing 8×12 culture arrays or 280 DPI for agar plates containing 16×24 culture arrays. The scanner comes with an accessory unit for scanning transmitted light images. Data collection is always performed using transmitted (film setting) light, equivalent to camera images with backlighting source. Concerning the PC, it was assumed that the commercial processor platform was sufficient to execute a Java program and save the results. Concerning agar plates, heated/liquid agar (like jelly) is poured into a plastic rectangular Petri dish with sidewalls, after it cools and/solidifies, yeast are spotted onto the surface. Light passes from above through the yeast cells, the agar, and the plastic bottom of the Petri dish and the glass platform of the scanning instrument to the detector.

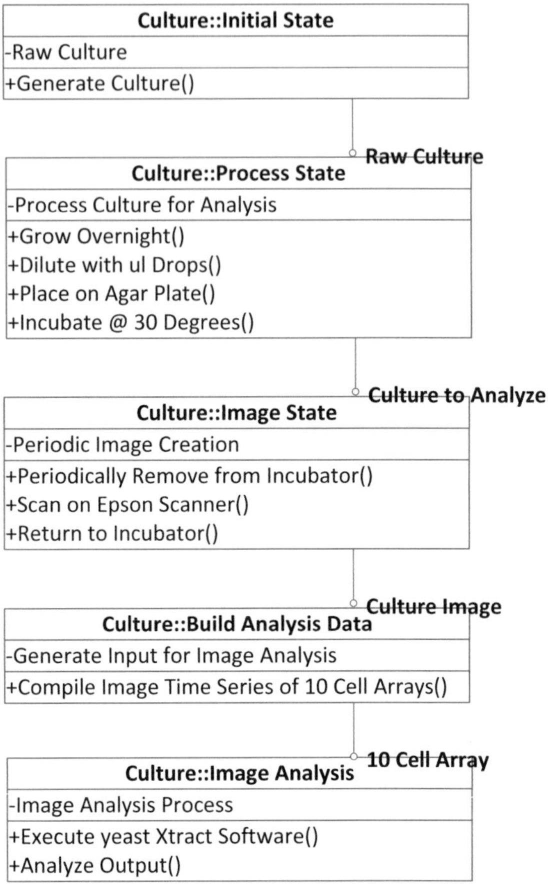

Fig. 2.9 System UML class diagram

A first look investigation of the method and device used to collect the imagery was performed. The basic collection method appears to experience normal image quality distortion problems in terms of focus, and lens distortion. The result of this distortion can cause variation in culture collects in terms of size and density based on the position of the culture spot near the center versus the edge of the scanner platform. These problems are inherent in any image collection that can be corrected by applying distortion correction mechanisms.

2.4.4 Current Software Environment

An analysis of the current Java software shows use of current Java libraries and Java2d only classes to perform automated analysis of the yeast experiments on cell array scans to determine whether changes have occurred and to what magnitude. This is accomplished by modeling an elliptical region slightly smaller than the yeast growth area and determining pixel recognition of values within the ellipse and outside the ellipse. Each "Culture Image" object holds the growth in pixels at a given moment in time. The author attempts to compensate for some of the error at the edges of each specimen by shifting pixels, et al., re-centering the ellipse and in all cases is summing up the intensity of all pixels for a later comparison against other specimens.

In the proposed example system changes below, we can show that with added algorithmic modifications for error correction due to scanner glass, etc. we can alter the software effectively. Additionally, we can then describe some overall software architectural changes to the system as a whole to speed analysis and data fusion in the existing lab, as well as, among external partners with data to share.

2.4.5 Current Experiment Algorithms

The program expects an initial input parameter file to set the scanner and other common parameters. Then it starts reading the input tiff file data that contains the image scans of ten agar plates containing 16 by 24 culture samples. Figure 2.10 [44] is an example of the input file.

Finally the software creates a temporary folder within a folder and copies this file over to temp. Then it is ready to start analysis processing. For each agar plate output the program creates a directory for each plate and then copies each annotated spot detection image into the proper directory. Figure 2.11 shows an example of an annotated spot detection file. Then as a final step the Main.java program creates three output ascii text files containing the spot intensities, and area for each culture spot by scan, row and column [44].

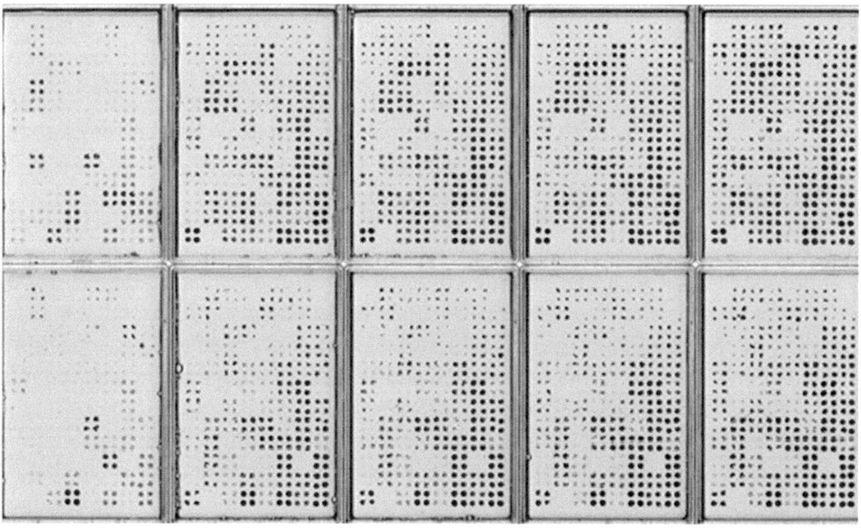

Fig. 2.10 Ten agar plate input scan

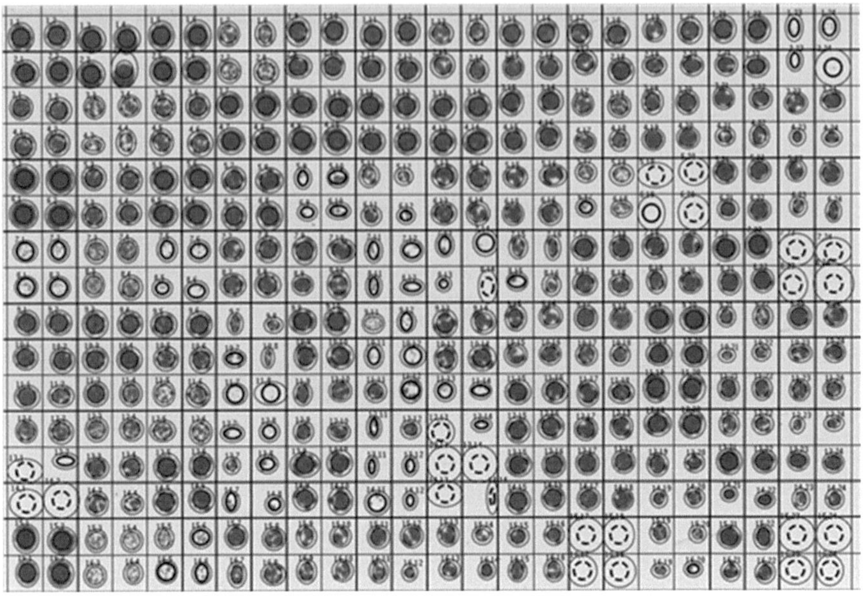

Fig. 2.11 Sample spot detection file

A Main.java function performs the major calculations. It considers background pixel content and figures out the ellipse and where the spot lies as an analog of the function it overlays. Then a function to detect the ellipse simply uses the values from the each set of spot passed-in values for all the other time points. Next the spots are aligned to the object's image so as to minimize the difference between it and an external image the reference image is moved accordingly, −x…+x and also −y…+y. Lastly, the software returns the average pixel intensity within each ellipse.

2.4.6 *System Upgrades*

2.4.6.1 Hardware and Infrastructure

The Systems Biology infrastructure field is changing from one of disjoint, sequential processes to an area that is centered on biological properties. These properties consist of models that incorporate sequence information, high throughput data, pathway models, etc. These biological properties are also scalable to data quantities [45]. We have realized through our analysis that causal error can be introduced into the system by the scanner, and agar plate. Specifically, the errors associated with the scanner are the fixed focal length, the linear distortion of the scanner lens and the radial distortion on the edge of the scanner lens. The agar plate error is comprised of the depth or thickness of Petri dish, agar gelatin constitution, and the reflectance of both. Due to these factors, there exists a density distortion factor of the agar.

2.4.6.2 Software Enhancements

Source Code Enhancements

Changes should be made to the algorithm as currently implemented to take into account the above mentioned error introduced by the scanner and the agar plate into the decision making process. The scanner focal length must be accounted for in the software as well as the radial distortion of the scanner lens. New algorithms should replace the current set that take into account summation of light and dark pixels to arrive at a value used to make the magnitude decisions. Hence, when implementing the lens modeling and algorithm implementations edge distortions should be removed from the equation or at least abated to a great degree.

Making the suggested modifications the magnitude output value will be much more exact and will also provide a value respective of the total volume of growth in three dimensions. The current implementation achieves only a 2D representation of magnitude. The Z-axis cell growth analysis encompasses potentially many more growth cells than perceived in 2D. Hence, potentially important to the classification and decision making process. Figure 2.12 illustrates this difference [41].

2.4.6.2.1 Software Architecture Enhancements

An analysis of the software architecture revealed that a stovepipe approach was used to develop the original system. For future scalability and system pliability we propose a potential re-architecting of the system involving implementing a true J2EE or otherwise known as Java Enterprise architecture. The following describes the components of a J2EE architecture template we will use in attacking the Lab architectural issues and use it to assist in organizing the lab components being analyzed [46]:

- Client Tier
  - Presentation layer

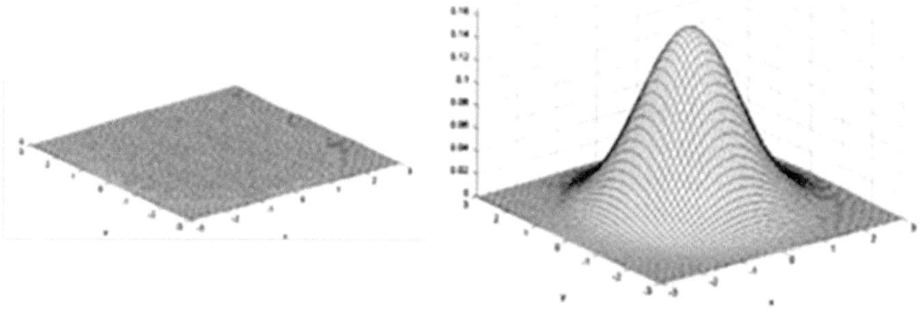

Fig. 2.12 2D vs. 3D sampling

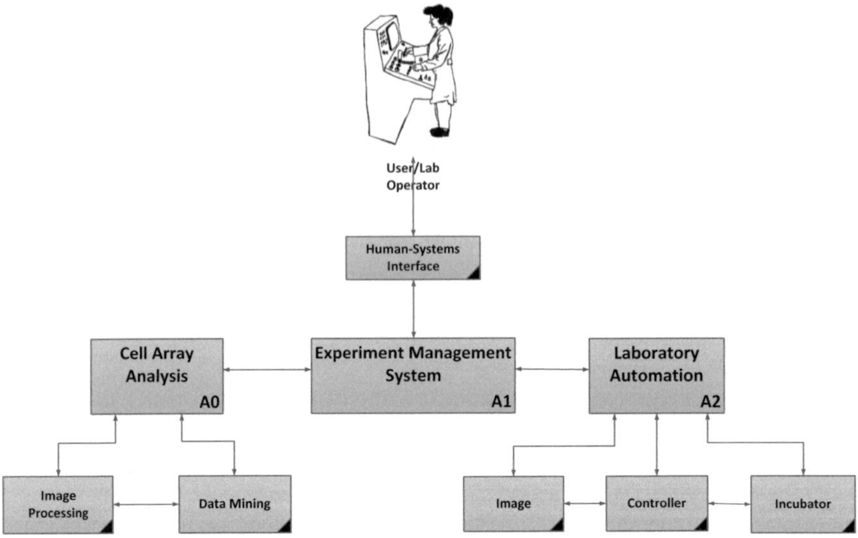

Fig. 2.13 Before J2EE

- Web Tier
  - Web Services
  - Web Applications
- Business Tier
  - Business Workflow Logic
- Enterprise Information System (EIS) Tier
  - Enterprise Data Tier

The design view of Fig. 2.13 illustrates the architecture before the J2EE enhancements, Fig. 2.14 presents a view of the architecture after the insertion of the J2EE technology and incorporates the following Architecture Tenets: Web-enabled, N-Tiered, Service Oriented (SOA), Component-based, Network-Centric, COTS-based, Collaborative Web GUI, Metadata Tagging, Seamless Data Access, Open Standards.

2.4.6.3 Algorithms Modifications

In order to relate changes in the real world to changes in the image collection sequences, we need parametric models that describe the real world environment and the image generation process. For this problem it is assumed that the real world environment is controlled and there is very limited number of variables that needs to be accommodated. In terms of image processing, the most important models the need to be considered are scene, object, camera, and illumination models. The YeastXtract program attempted to account for scene (background) and object

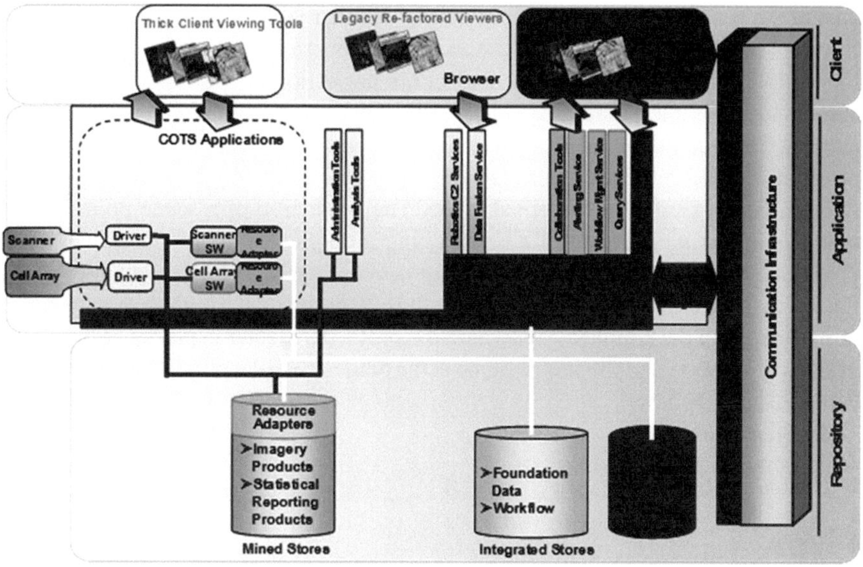

Fig. 2.14 After J2EE

Table 2.2 Naming conventions for real-world model entities

Real world	Parametric model	Model world
Real scene	Scene model	Model scene
Real object	Object model	Model object
Real texture	Texture model	Model texture
Real shape	Shape model	Model shape
Real camera	Camera model	Model camera
Real image	Image model	Model image
Real illumination	Illumination model	Model illumination

(culture spot) but did not provide any consideration adjustment in camera or illumination. Further, by using a scanner vice a digital camera (pinhole model solution) then the creation of a parametric model to compensate for collector distortion becomes much more difficult. Table 2.2 shows the terms that we use to name the parametric models, their corresponding real world entities, and the entities reconstructed according to the parametric model and presented in the real world [47].

At this point each model is covered starting with the camera model followed by illumination, object and finally scene.

2.4.7 Camera Model

The camera model describes the projection of real objects in the real scene onto the image plane of the real camera. The imaging plane is also referred to as the camera target. The image on the image plane is then converted into a digital image.

Due to the proprietary nature of commercial scanner venders, instead of a scanner, if a high quality digital camera was used then camera modeling becomes straight forward. A widely used approximation of the projection of real objects onto a real camera target is the pinhole camera model. This corresponds to an ideal pinhole camera. The geometric process for image formation in a pinhole camera has been nicely illustrated by. The process is completely determined by choosing a perspective projection center and a retinal plane. The projection of a scene point is then obtained as the intersection of a line passing through this point and the center of projection C with the retinal plane R. Most cameras are described relatively well by this model. In some cases additional effects (e.g. radial distortion) have to be taken into account. In the simplest case where the projection center is placed at the origin of the world frame and the image plane is the plane Z = 1, the projection process can be modeled as follows:

Image Projection

$$x = \frac{X}{Z} \quad y = \frac{Y}{Z} \tag{2.1}$$

For a world point (X,Y,Z) and the corresponding image point (x,y). Using the homogeneous representation of the points a linear projection equation is obtained:

$$\begin{bmatrix} x \\ y \\ 1 \end{bmatrix} \sim \begin{bmatrix} 1 & 0 & 0 & 0 \\ 0 & 1 & 0 & 0 \\ 0 & 0 & 1 & 0 \end{bmatrix} \begin{bmatrix} X \\ Y \\ Z \\ 1 \end{bmatrix}$$

This projection is illustrated in Fig. 2.15. The optical axis passes through the center of projection C and is orthogonal to the retinal plane R. Its intersection with the retinal plane is defined as the principal point c.

With an actual camera the focal length f (i.e. the distance between the center of projection and the retinal plane) will be different from 1; the coordinates of (2.1) should therefore be scaled with f to take this into account. In addition the coordinates in the image do not correspond to the physical coordinates in the retinal plane. With a CCD camera the relation between both depends on the size and shape of the pixels and of the position of the CCD chip in the camera. With a standard photo camera it depends on the scanning process through which the images are digitized. The transformation is illustrated in Fig. 2.15.

The image coordinates are obtained through the following equations

Image Transformation

$$\begin{bmatrix} x \\ y \\ 1 \end{bmatrix} = \begin{bmatrix} \frac{f}{p_x} & (\tan\alpha)\frac{f}{p_y} & c_x \\ & \frac{f}{p_y} & c_y \\ & & 1 \end{bmatrix} \begin{bmatrix} x_R \\ y_R \\ 1 \end{bmatrix} \tag{2.2}$$

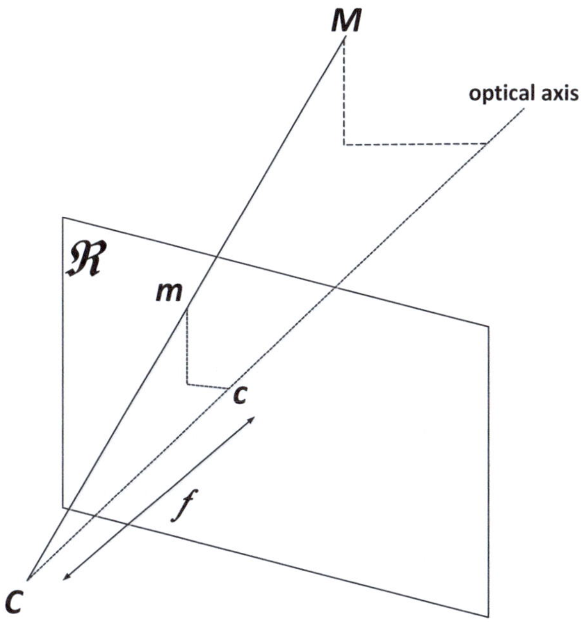

Fig. 2.15 Perspective projection

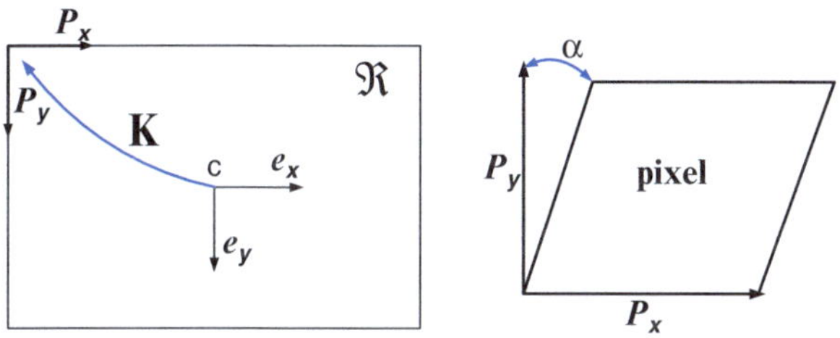

Fig. 2.16 Retinal coordinates to image coordinates

where p_x and p_y are the width and the height of the pixels, T is the principal point and α the skew angle as indicated in Fig. 2.16. Since only the ratios and are of importance the simplified notations of the following equation will be used in the remainder of this text:

Simplified Equation of (2.2)

$$\begin{bmatrix} x \\ y \\ 1 \end{bmatrix} = \begin{bmatrix} f_x & s & c_x \\ & f_y & c_y \\ & & 1 \end{bmatrix} \begin{bmatrix} x_R \\ y_R \\ 1 \end{bmatrix} \tag{2.3}$$

with and being the focal length measured in width and height of the pixels, and a factor accounting for the skew due to non-rectangular pixels. The above upper triangular matrix is called the calibration matrix of the camera; and the notation will be used for it. So, the following equation describes the transformation from retinal coordinates to image coordinates.

$$m = Km_{\Re}$$

For most cameras the pixels are almost perfectly rectangular, and thus are very close to zero. Furthermore, the principal point is often close to the center of the image. These assumptions can often be used, certainly to get a suitable initialization for more complex iterative estimation procedures. For a camera with fixed optics these parameters are identical for all the images taken with the camera. For cameras which have zooming and focusing capabilities the focal length can obviously change, but also the principal point can vary. An extensive discussion of this subject can for example be found in the work of Willson [48]. At this point it is unclear which type of camera Dr. Hartman's laboratory team may migrate to in the future.

2.4.8 Illumination Model

There are two types of light sources: illuminating and reflecting. An illumination model describes how the light incident on the object influences the reflected light distribution, which is what we see. There are several such models. Illumination models can be divided into spectral and geometric models. Spectral illumination models are used if we want to model changes of color resulting from several colored light sources, indirect illumination with objects of different colors, or colored reflecting surfaces. Geometric models describe the amplitude and directional distribution of incident light. Geometric illumination models are available for ambient and for point light sources. The problem covered in this paper centers around a geometric illumination model for a point light source projecting through a semi-transparent agar background. Djokic has proposed an efficient analytical procedure for using an extended ideally diffuse light source (IDLS) with one point light source [49]. He provides example luminance calculations (horizontal, vertical and vector/magnitude) for a rectangular IDLS. Some aspects of the proposed procedure are considered in terms of photometric measurements and practical applications [49]. Hank Weghorst describes a set of algorithmic procedures that have been implemented to reduce the computational expense of producing ray-traced image models [50]. Concerning the semi-transparent agar background, YeastXtract program calculates the value of the semi-transparent background for boundary areas containing each yeast culture spot it is unclear where that value is used to adjust the calculate gray scale value associated with the yeast culture spot. During data investigation we calculated the background across all sample data files and determined that the average agar color value 225.4 on the gray scale (0 = max black to 255 = max white). Further the lowest agar color value was 222 and the highest value was 228.

This implies two things. First the light color frequency provides a gray scale value greater than 228. Second, the light color values below the gray scale value 221 belongs to yeast culture spots. Based on our data and source code inspection, it is recommended that the background value be calculated once, set in a value, and then applied it to the yeast culture spot. Pixel density values vice being calculate over and over for boundary areas containing each yeast culture spot.

2.4.9 *Object Model*

The object model describes assumptions about the real objects. In this problem the objects are yeast cultures in different phases of maturation. Texture models describe the surface properties of the object. In yeast cultures, if a yeast culture can exhibit different textural characteristics then these textural characteristics must be considered for inclusion in the textural components of the object model. Here, we assume that the texture of each yeast culture point is described by a color parameter. The YeastXtract program calculates the gray value for each object pixel belonging to each yeast culture spot and calculates and average gray scale value. The YeastXtract program equates each average gray scale value as a density value. To obtain a real density value for each yeast culture spot area one should create a gray scale to micron thickness value table for the valid range of gray scale (estimated to be gray scale values between 100 and 222) than modify the YeastXtract program to calculate the actual dimensions for each pixel belonging to the each yeast culture spot over time.

2.4.10 *Scene Model*

Having discussed how to model the illumination source, object, and camera, now the focus is on modeling the image scene. The scene model describes the world with illumination sources, objects, and cameras. Depending on the models used for different components, one arrives at different scene model. Since there should be one simple camera, illumination, and object modal the scene model is just applying the sum of the corrections of the three models discuss above.

2.4.11 *Example of Discovery Research for Systems Biology Infrastructure*

Current work in Systems Biology provides information on an up and coming new research paradigm. Advances in the biology discipline as DNA sequencing, the Genome project are key new areas in this field. These advances have changed the research paradigm that had been used in biology. A key attribute of this new research paradigm are the transdisciplinary research teams that work together to design experiments, collect and process data for these focused biological areas termed "-omics". Examples of this distinct research areas are genomics (study of whole genomes) and proteomics (study of all proteins an organism makes).

Systems Biology utilizes a computing infrastructure designed to apply data management and analysis techniques to process data from areas such as gene/ protein clusters to their existing functional context recorded in public databases such as GenBank, KEGG, GO and OMIM. This area within Systems Biology is specifically called Bioinformatics. The influx of data is exploding with new types of data and large amounts of the currently collected data. New techniques are being developed to integrate, manage, interpret and visualize data from diverse sources to continue advances in the Biology discipline.

In the past, the main focus of the Bioinformatics field is to develop algorithms to analyze genomic sequences in massive databases [51]. Advances in this field have necessitated a Bioinformatics infrastructure change from an analytical interpretation of data to one that deals with biological properties. These biological properties consist of cross domain models of biological processes that incorporate "sequence information, high throughput data, ontological annotation, pathway models and literature sources." This better allows the cross discipline teams' access to all of the data in a form that is meaningful to their particular areas of expertise. It has been brought about by the need to have data and application integration on diverse execution platforms. The technology base for this is current state of the art for service oriented architecture and web services. Cheng provides a succinct goal for Systems Biology infrastructure: "Infrastructure's primary task is to reverse engineer biological circuits [24]. But we also expect it to apply to bioengineering projects by supporting the design and synthesis of complex sets of molecular interactions with a particular computational, biomedical or biosynthetic purpose [52].

The initial discussion centers on two major areas of concern related to infrastructure requirements. Data integration as related to database and application integration is the first requirement discussed. Both require the retrieval and storage of data distributed across heterogeneous local and public data sources. Both relate to a single key concept. Data is unstructured and mainly text fields. Sources for this data are on the increase. The data is continually being analyzed and cross-referenced. These new outputs are then stored in new organism specific, domain specific or disease specific repositories. DiscoveryNet efforts resolved this disparity issue by using a uniform interface for queries that span relational and XML repositories. It is based on a wrapper concept that allows selection of data widgets which are customized to each of the various repository formats.

The second issue relates to application integration. A survey conducted with Biology professionals performed for DiscoveryNet resulted in an efficiency workflow concept. In other words, the need is to provide a workflow logic that will run independent tasks in parallel and dependent tasks in the correct sequence. The former will also necessitate combining the results of these parallel tasks at the end of the process. Currently, tools to meet these requirements are being developed by researchers and made available over the web. As these tools proliferate, a rapid integration framework is necessary to enable the inclusion of these new tools as they are needed by the researchers. The functional elements of DiscoveryNet are primarily geared toward supplementing the knowledge discovery process for the Systems Biology discipline. This process is based on defining architecture for the explicit and implicit application pipelines that provide the complex analysis associated with Systems Biology. It is extended to include external applications along with the many data repositories utilized in the analysis tasks. These include the following:

- Component platform supporting deployment of functional components
- Execution management that includes Meta information to control parallel task execution and serial task execution in the proper sequence.
- Data modeling to represent the data structures that are being passed between the components
- Component registry that will make these components available to the research community in general.
- Deployment issues such as location, instantiation, etc.
- Collaboration support for the design and execution of components.
- Service engine treating the components as available services for the research community.

The field of Systems Biology focuses on analysis of molecular systems [53]. The infrastructure provides computational support to the biologists and other scientists who are performing the analysis for this data. That basic key roadmap for a systems biology project identifies four steps:

- Identify all of the data types that are to be handled by the system (genes, proteins, etc.)
- Provide high throughput technologies such as micro-arrays used for analysis of complex biological systems
- Generate a global model that provides a biological, mathematical and computational representation of the complex biological system.
- Model analysis to provide a new hypothesis for testing.

A key concept in building an infrastructure such as this is scalability. Each component needs to be scalable in its biological objects, processes and intermediate interfaces amongst the components [53].

2.4.12 Multidisciplinary Application Discussion

Justification of Multidisciplinary processes used to augment Systems Biology and the yeast processing lab have been detailed above. The existing system was analyzed using a multi-tiered architectural approach used in software engineering. The Presentation Tier described the possible visual aids for the user, while the Application Tier described modifications and adaptations to the existing system (algorithms, software) which enhance the quality and efficiency of generating and sharing the data. Additionally, a future enhancement should be to develop a specific camera model to handle all of the data collection errors described above. The data layer was analyzed for standardization to enhance the sharing of the data [48].

Further work is needed in the area of accurate, precise modeling of cell proliferation kinetics from time-lapse imaging and automated image analysis of agar yeast culture arrays [54]. One area is in data collection. There is a need to create a method to correct agar plate data collections for environment induced errors due to collector and other environment conditions. This could be achieved by creating an imagery distortion correction processing algorithm and associate calibration parameter table in Java to be use in performing image correction on the data prior to quantitative systems level analysis processing of gene interactions. A second area is in creating a more rigorous experiment setup process. A third area is in the automated image analysis processing itself. Currently the software program is a complex compute intensive program that can be restructured and streamlined and the precise processing is performing very fine calculations on very crudely collected data.

2.5 Discussion

Modern systems require modern thinking. Multidisciplinary Systems Engineering provides a new approach and paradigm for Systems Engineering organizations; a necessary shift to take System of Systems design into the future. Even though many organizations recognize that Systems Engineering is inherently a multidisciplinary approach, most do not really embrace the concept and train their engineers to think along multidisciplinary lines. Each of the authors, having spent decades designing, architecting, implementing, and delivering large-scale system have come to appreciate the need for this book. The case study provided in Chap. 2 illustrated one example of the multidisciplinary approach as applied the study of Biology. MDSE's emphasis on bringing all disciplines to the table at the same time holds the promise of more robust System of Systems architectures, holding the possibilities of improvement in achieving overall goals and constraints placed on modern system designs and a reduction in the variability created by designing system disciplines separately. Recognizing that MDSE is inherently a feedback driven process, and not being pushed by "just get it done" attitudes will, in the end, produce architectures that better meet the demands of ever more challenging system designs.

Question to think about:
Is there any inherit difference in MDSE between a traditional information management system and the Systems Biology problem?

Chapter 3
Multidisciplinary Systems Engineering Roles

Multidisciplinary System Engineer's play many roles within the scope of the entire System of Systems lifecycle. In some cases these roles run throughout the program. In some cases Multidisciplinary Systems Engineers take on multiple roles at different stages in the System of Systems development. Here we describe the basic roles of System Engineering within a program and how they are involved in every aspect of the overall system lifecycle.

> ***Question to think about:***
> *What actually happens in a system development when Systems Engineering does not define the Service Level Agreements on exposed system services?*

3.1 Systems Architect/Analyst

Systems Engineering (SE) owns, manages, allocates, writes, and maintains requirements for the program. Systems Engineering utilizes these requirements to provide functional decomposition/allocation and development of the overall systems architecture, including derivation of sub-system requirements from the customer system specification. This includes derivation of service definitions and Service Level Agreements (SLA)[1] to be utilized within the overall system.

The role of a SE Architect/Analyst is a dual hat role, where Analyst and Architect are used interchangeably in the system engineering field today. In some cases, an organization may utilize both an Analysis and Architect as independent roles, but in

[1] A Service-Level Agreement is part of Web Service contract, where the Web Service is formally defined in terms of the Web Service scope, quality, and overall service responsibilities; this agreement is made between the service provider and service user.

J.A. Crowder et al., *Multidisciplinary Systems Engineering*,
DOI 10.1007/978-3-319-22398-8_3

general, the Analysis and Architect are one and the same role. This role encompasses the translation of customer development, operational, and supportability needs and system requirements into well-written operational/technical requirements for which system and subsystems (and services) can be architected and designed (e.g., hardware and software). This requirements decomposition includes understanding all internal and external interfaces and ensuring the functional architecture correctly and completely captures the customer's needs.

As potential changes to the system are generated, it is the role of SE architect to assess the impacts to the overall system and to the subsystems, based on which specific requirements must be modified. The Systems Architect is responsible for architectural designs for projects that address mission/business, application, information, and infrastructure requirements. This role is crucial to the successful delivery of a system that meets the technical requirements and standards, and provides support throughout the development life cycle with support to the mission/business analyst, System Engineering (including Infrastructure). The System Architect/Analyst supports pre-implementation, development/implementation, and post implementation (test and delivery) of the overall system. The role of Systems Architect/Analyst focuses on:

1. Defining the business/mission needs for the system.
2. Defining the solution options, the technology gaps, and technology recommendations.
3. Driving the solution evaluation and decision selection, including buy vs. build, and early project estimates (high-level) in terms of cost, schedule, and capabilities.
4. Ensuring that alternative solutions meet technical requirements, as well as quality attributes for the system, like robustness, availability, reliability, extensibility, etc.
5. Ensuring that the architecture and design complies with existing industry standards, corporate technology roadmaps, compliance documents (e.g., security, FAA, OSHA, etc.), and enforce systems and software industry standards.
6. Participating in architecture assessments[2] in order to drive out potential risks that may inhibit the successful implementation or performance of the system under development.
7. Guiding and supporting other system architects (including COTS vendors and subcontractors) in the design and development of the Information Technology and Information Systems architectures.
8. Oversight of all functional, technical and performance test execution.

[2] Typically utilizing the Architecture Tradeoff Analysis Method (ATAM); an industry methodology for evaluating systems and software architectures relative to the overall specified system quality attributes.

Three very important things to keep in mind about the role of a Systems Architect/Analyst:

1. A more complicated architecture does not make it a good architecture. A really good architecture meets the customer requirements, does not violate established practices and principles of System Engineering and Architecture, takes into account performance and quality attributes, is acceptable to the user, and, is as simple as possible to meet these architecture quality characteristics.
2. System Architecture design is an iterative process. Many interface interactions and simplifications only become apparent as the architecture is decomposed. This requires the System Architect/Analyst to iterate on previous, or higher-level, architecture layers to see if a different overall decomposition would simplify or clarify lower architecture layers.
3. Risk is inherent in Systems Architecture. Every system has risks associated with each of the different implementation architectures. Risks cannot be completely eliminated, but can often be contained in terms of their probability of occurrence and/or overall impact to the system. A well designed architecture should be robust, and include contingent solutions that allow the development or operation of the system to be maintained, if the potential risks materialize.

Question to think about:
What happens in a system development when the System Architect Role is staffed by a Software Architect?

3.1.1 The Architecture Tradeoff Analysis Method (ATAM)

It is critically important to ensure that the designed system and software architectures for a given development effort meets the requirements, constraints, performance, and quality attributes that define the customer and user needs of a system. Of the analysis methods that are available for evaluating systems and software architectures, one commonly used method is the Architecture Tradeoff Analysis Method (ATAM) [55]. The purpose the ATAM is expose potential risks that may cause the design, once implemented, to fail to meet the overall mission/business goals specified by the customer. The ATAM assesses the architectural design through analysis of the Quality Attributes for the proposed architecture. One important note about the ATAM, it is not an indication of the quality of the design, only the risks associated with the consequences or the architectural decisions. Performing the ATAM will, however, force preparation, documentation, and understanding of the proposed systems and software architectures. The Quality Attributes include, but are not limited to:

The Mission/Business Quality Attributes:

- Schedule—Time to Market
- Cost and Benefits to users

- Life expectancy of system use
- Targeted Market—who is the system for?
- Legacy System Integration
- Roll back potential and issues

The User's Quality Attributes:

- Performance
- Availability
- Reliability
- Maintainability
- Usability
- Security

The Developer's Quality Attributes:

- Portability
- Reusability
- Testability

Figure 3.1 illustrates the overall ATAM process. The User Scenarios or Use Cases, discussed in Chap. 4, are utilized to represent the customers' interests and to understand their perspective on the quality attributes to be analyzed. They provide the way the system is anticipated to be used across several different scenarios or mission/business needs, including introducing faults and new stimuli into the system to show how the system will respond to unanticipated stresses.

The Sensitivity Analysis shown in Fig. 3.1 provides a measure of how parameters in the architecture affect various Quality Attributes. The Trade-Off Analysis

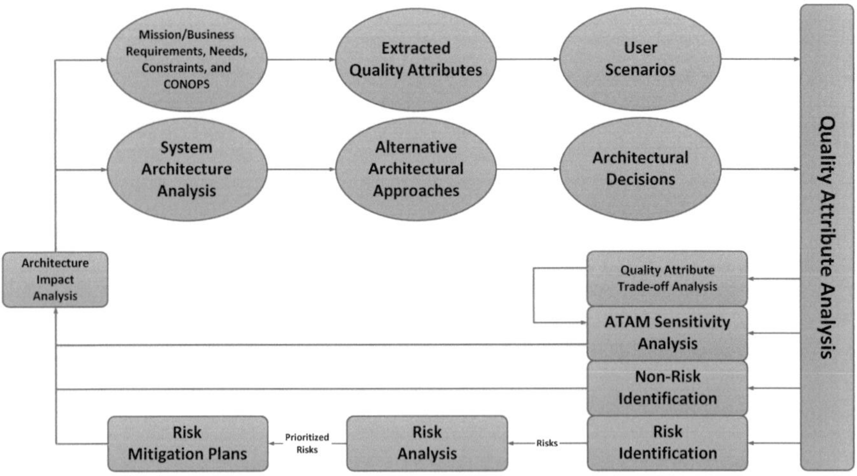

Fig. 3.1 High-level ATAM process

determines if architectural parameters affect more than one Quality Attribute. An example would be if increasing desired usability in a certain way makes the system less secure. There would have to be a trade-off analysis between these two quality attributes to determine which is more desirable in the overall system design. Remember, the ATAM is one tool in the System Engineering arsenal to help in assessing the overall applicability of a given architectural approach to the system to be developed. There is no single analysis that assesses the overall quality of the Systems Architecture. The Systems Engineering Architect/Analyst must apply those analyses that make sense for a given system, its size, complexity, and overall mission/business criticality.

Questions to think about:
Does an ATAM really count if all they do is evaluate compliance to metrics? What must be done in an ATAM to ensure it provides the proper assessment?

3.1.1.1 Quality Functional Deployment

Quality Functional Deployment (QFD) is another tool that can be used to determine the applicability and provides a rational basis for proposed architectural concepts and proposed system design for a given program. The QFD process consists of seven different tools [58]:

- Affinity Diagrams: used for brainstorming
- Interrelationship digraphs: used to make sure the design problems are well understood
- Tree Diagrams: used the hierarchically structure design ideas and system components
- Matrix Diagrams: used to correlate the tree diagrams
- Matrix Data Analysis: used to analyze the tree diagrams
- Decision Charts: used to capture system failure modes
- Arrow Diagrams: used to map out the final architecture design process

Unlike development efforts of the past, where change was regarded as risk, in modern programs and system development change has become an integral part of the overall execution flow throughout each stage of the development process. Agile development has become the norm, even though Systems Engineering has been slow to keep up [56]. In this change-driven environment that strives for continual improvement, QFD provides the necessary rational for these changes. Providing a methodology and discipline for integrating the customers' needs, and possible changing requirements, it enables the design/development team the methods to evaluate and integrate these changes into a solid design change vision.

QFD also provides a solid, rational basis for identifying new technologies and concepts that are applicable to the system development throughout its lifecycle.

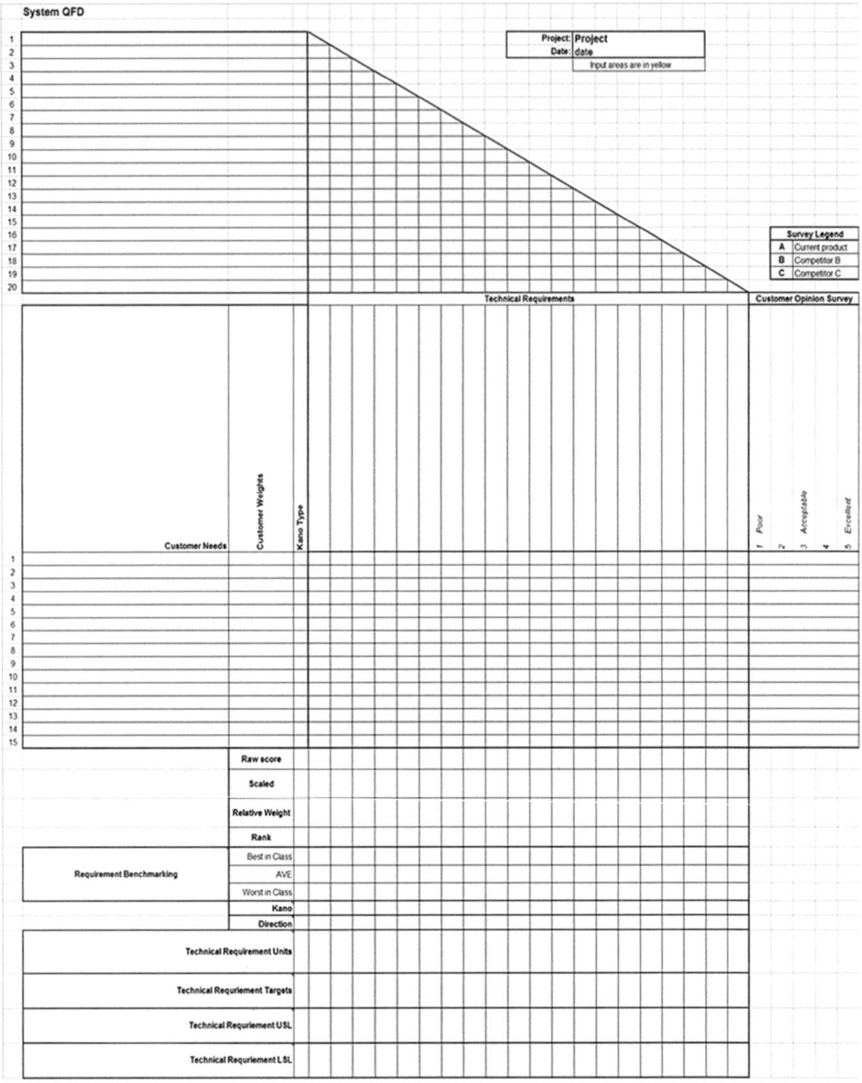

Fig. 3.2 QFD A1 quality matrix

If properly utilized an a systems design tool, QFD gives the architecture, analysis, and design engineers traceability from initial requirements analysis to the final product design, before implementation and/or manufacturing have begun [57]. Another aspect of the QFD process is that it allows Systems Engineering to manage the overall objectives and drives the desired design robustness characteristics (quality attributes) into the system and software architecture. The use of QFD method and Multidisciplinary System Engineering is a paradigm shift from the sequential

Table 3.1 The A3 quality attribute cross-correlation matric

	Reliability	Maintain-ability	Flexibility	Net centricity	Scalability	Integrity	Availability
Reliability	X						X
Maintainability		X					
Flexibility			X		X		
Net centricity			X	X			
Scalability			X		X		
Integrity	X					X	
Availability							X

Table 3.2 SoS enterprise fault error sources

Source of error	Frequency (%)
Requirements	8.1
Features and functionality	16.2
Structural bugs	25.2
Data bugs	22.4
Implementation and coding	9.9
Integration	9.0
System and software architecture	4.7
Test definition and execution	2.8
Other	1.7

Source: Boardman, J., Dimario, M., Sauser, B, and Verma, D. 2006. System of Systems Characteristics and Interoperability in Joint Command and Control. Defense Acquisition University

development approach to an agile, change embracing approach required for modern system design and development. The agile Systems Engineering approach utilizes the multidisciplinary System Engineering team to pre-empt design faults, especially when the QFD process is utilized.

The QFD Matrix methodology consists of several matrices, at different levels of detail for the architecture. The first three are important for hardware and software (service) management. An example of the A1 Matrix is shown in Fig. 3.2 and the A3 Matrix is shown in Table 3.1. The fourth (A4) matrix develops the next level of detail for the design, identifying the system hardware and software components required for the System Architecture design and implementation. An example of the A4 Matrix is shown in Fig. 3.3.

Other matrices are created to evaluate new concept and new technologies that may be appropriate for the systems. Each of the Quality Attribute, Functionality, Architecture Component matrices and trees are evaluated independently against new technology and concept ideas. The combination of these matrices forms the basis for a quality architecture design and an information database to be used throughout the development lifecycle. Figure 3.4 illustrates QFD within the continuous process/product improvement development lifecycle [58].

Subsystem QFD		Subsystem Requirements																			
Technical Requirements	Tech Rqmt Relative Weights																				
	Raw score																				
	Relative Importance	0%	0%	0%	0%	0%	0%	0%	0%	0%	0%	0%	0%	0%	0%	0%	0%	0%	0%	0%	0%
	Rank																				

Fig. 3.3 The A4 QFD architecture domain matrix

3.2 System Designer

The System Designer takes into account all of the variables in the system that need to be created, based on the proposed architecture, including the hardware and software required to facilitate the architecture, the data and information, how it travels and is transformed throughout the proposed architecture. This is an iterative process, since issues discovered in this phase may require a change to the architecture. This role comes into play after the requirements and functional decomposition has been completed. In reality, these two roles overlap and require much interaction, particularly in the selection of subsystems and development of subsystem requirements that lead to seamless overall integration of the system.

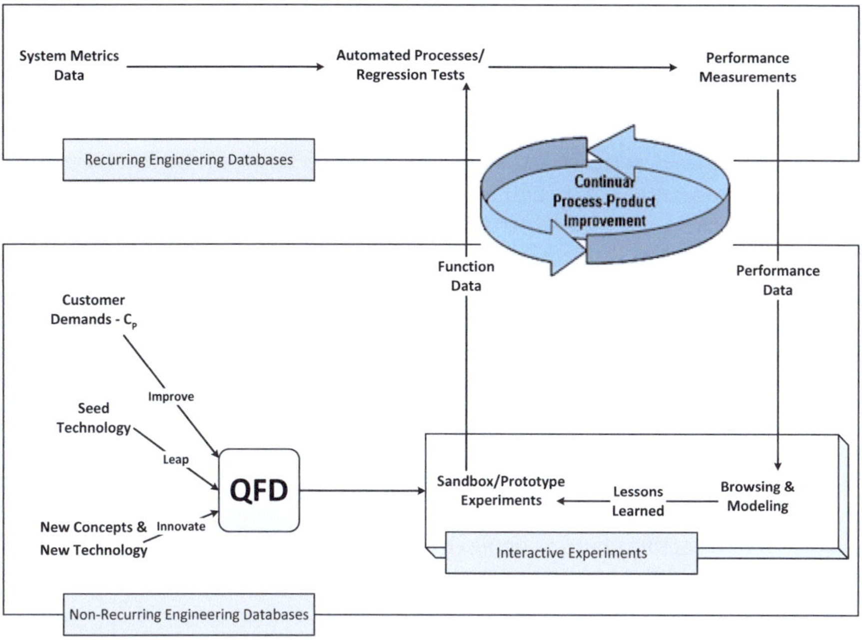

Fig. 3.4 QFD as part of a continuous process/product improvement lifecycle

In this role, the systems engineer(s) work with the Systems Architect/Analyst to define the components, modules, data, and interfaces (both external and internal) required for the system to satisfy the prescribed requirements. The Systems Designer's role is to take the theoretical proposed System Architecture and apply it to product development. There are two main approaches to Systems Design, top-down and bottoms-up.

3.2.1 Top-Down Design Methodology

A **top-down** approach (also known *decomposition*) is essentially the breaking down of a system to gain insight into its compositional elements and sub-systems. In a top-down approach an overview of the system is formulated, specifying but not detailing any first-level subsystems. Each subsystem is then refined in yet greater detail, sometimes in many additional subsystem levels, until the entire specification is reduced to base elements. A top-down model is often specified with the assistance of "black boxes", these make it easier to manipulate. However, black boxes may fail to elucidate elementary mechanisms or be detailed enough to realistically validate the model and must be decomposed further in order to understand the required lower-level functionality. Top down approach starts with the big picture and then breaks it down from there into smaller segments.

3.2.2 *Bottoms-Up Design Methodology*

A **bottom-up** approach is the piecing together of low-level parts of the system to give rise to more complex parts of the system and sub-systems, thus making the original lower-level systems and sub-systems of the emergent high-level system. Bottom-up processing is a type of information processing based on incoming data from the environment to form a perception. Information is gathered by the low-level architects (input), and is then turned into a concept that can be interpreted and recognized as new low-level functions (output).

In a bottom-up approach the individual base elements of the system are first specified in great detail. These elements are then linked together to form larger subsystems, which then in turn are linked, sometimes in many levels, until a complete top-level system is formed. This strategy often resembles a "seed" model, whereby the beginnings are small but eventually grow in complexity and completeness. However, this constitutes what could be called "organic strategies" and will result in a tangle of elements and subsystems, developed in isolation, and are most often subject to local optimization as opposed to meeting the global (System of Systems Level) purposes [59].

Question to think about:
When bottoms-up engineering is used on a System of Systems solution, what are the possible negative effects to the project?

3.3 System Integrator

The role of a System Integrator is to bring together the various subsystems, components, and services for the program and brings them together to work as a complete system. The System Integrator may also be involved in bringing together a number of legacy systems, or one or more legacy systems and a new system under development. This includes the design and integration of legacy and new program controls, external and internal interfaces, data transformations, and communications protocols.

The System Integrator must also ensure that the physical implementation of the system have been thoroughly addressed by the systems and software architects. The System Integrator must determine whether the integrated system will meet the mission/business goals, requirements, needs, and constraints of the system customer(s). This must include the physical, logical, and environmental aspects of the system function together as a complete system.

The Systems Integrator must bring a wide range of skills, demonstrating a breadth of knowledge in systems engineering, systems and software architecture, hardware and software engineering, as well as interface protocols. They need to be good problem solvers, in that every systems integration effort produces new and challenging issues that must be worked through in order for the completed system to operate as intended.

There are three main integration processes, called Vertical, Horizontal, and Star Integration. It is the role of the Systems Integrator to analyze and identify the challenges for system integration, and apply one or more of the integration process during system integration.

3.3.1 Vertical Integration

Vertical Integration involves integrating subsystems, enterprise services and components according to a logical grouping of functionality. Functional entities are created, sometimes called Functional Integration Silos or “stovepipes”. These silos provide a less expensive and shorter term integration effort than Horizontal Integration, but there is no guarantee that these silos will work together to form a complete, cohesive, functioning system. Because of this, the total life-cycle cost can be substantially higher when rework is needed when silos do not function well with other silos.

In addition, scaling the system to add capabilities is often only facilitated through the implementation of other silos. Reusing subsystems is rarely possible in this case. In short, this makes Vertical Integration complex, expensive, and inherently risky. Unfortunately, many programs opt for this methodology because of its up-front reduced cost and short time frame. Rarely does Vertical Integration turn out to be the right strategy, particularly for modern systems.

3.3.2 Horizontal Integration

Horizontal Integration is an integration methodology in which subsystems, enterprise services, and components are required to communicate with other subsystems, which allow the Systems Integrator to identify the internal and/or external interfaces required by each subsystem, service, and component; working to minimize the interfaces to the Enterprise Infrastructure. One implementation methodology to accomplish Horizontal Integration is through the use of an Enterprise Service Bus.[3]

The Enterprise Service Bus allows a single interface for each subsystem, enterprise service, or component, providing translation capabilities between each interface. The definition of the interfaces is where the Systems Integrator can exert control over the final system implementation, as it is the job of the system integrator to ensure the interfaces operate properly Utilizing the Horizontal Integration methodology and decoupling capabilities provided through the service bus, the Systems Integrator has flexibility, in that it is possible to completely replace one subsystem with another subsystem using the service bus and bridging services, where the new capabilities are isolated and become transparent to the rest of the subsystems.

[3] An Enterprise Service Bus (ESB) is a software model (usually implemented using COTS S/W) used for implementing communication between interacting software services.

3.3.3 Star Integration

In Star Integration, each element of the System of Systems subsystems is connected to each of the other subsystems; hence the system integration diagram resembles a star. This drives interfaces between each subsystem, which can drive up integration costs, particularly as the number of subsystems increases; since the protocols and information interfaces may be different.

> ***Question to think about:***
> *What roles must work together to enable the System of Systems to integrate?*

3.4 Systems Information Security Engineer

The Information Security Engineer works with the system architects, designers, and implementers to provide system designs that accurately mediate and enforce the security policies defined in the customer requirements and compliance documents. This includes all aspects of system design and implementation, including theory, architecture design, testing, software engineering, and IV&V. Information Security engineers work through all phases of the development to ensure system availability, information integrity, user/process authentication, confidentiality, and non-repudiation [60]. These engineers handle any information assurance components of the system design, which endeavors to keep data/information secure and involves technologies such as cryptography.[4] Systems Information Security Engineers must involve many disciplines to provide a complete look at system security. This includes, but is not limited to, psychology, social science, fault tree analysis, and software engineering in the pursuit of ensuring the overall system meets all of its security requirements and constraints.

> ***Question to think about:***
> *What happens to the risk posture of a system when Information Security is not designed in with a holistic requirement set?*

3.5 Configuration Management

Configuration Management (CM) includes document, data, and metrics management. Management and control of documents, information, and metrics is vital to ensuring that the correct versions of information/documents/data/metrics are used

[4] Cryptography involves technologies for secure communications between two parties or systems.

throughout the program. Any changes to any Configuration Controlled source must go through a formal process to make a change, where changes are documented and then distributed throughout the project. While this is typically a separate group from System Engineering, there must be tight coordination between CM and System Engineering to ensure all plans, procedures, and architectures are follow the CM process. Failures to create, utilize, and maintain a program-wide configuration management system can have catastrophic results; as the following Case Study describes.

> ***Question to think about:***
> *Even with formal Configuration Management processes and practices, does Configuration Management existence actually ensure the program quality is maintained?*

3.6 Case Study: Knowledge Management and the Need for Formalization

Knowledge management is an important aspect of the MDSE formalism that must be accounted for in any System of Systems design. Knowledge management inherently involves both Data and Information Management throughout each element, subsystem, CI, and service-service interaction. One common way to capture the data/information flows throughout the system and, at a high level, capture System of Systems Knowledge Management is through the use of Ontologies. First we will introduce the notion of Ontologies and Taxonomies and then we will illustrate an example of a System of Systems Enterprise Fault Ontology.

3.6.1 Ontology Development

Ontologies are an important part of conceptual modeling needed for System of Systems Architectures. They provide substantial structural information, and are typically the key elements in any large-scale integration effort. Here we discuss several notions from the formal practice of ontology, and adapt them for use in System of Systems analysis as it applies to Knowledge Management; a follow on the Data/Information Management [61]. The aim is to provide a solid logical framework that will allow formal taxonomies for the elements of a System of Systems to be analyzed. The purpose here is to discuss some issues with respect to Knowledge Management ontologies and taxonomies for System of Systems development. A notional ontology for Knowledge Management is illustrated, where we look at what issues exist for ontology-based Knowledge Management systems in a large, diverse System of Systems Environment.

3.6.1.1 The Need for Ontology in a System of Systems Environment

With the ever-increasing availability of data/information that must be processes by large System of Systems programs (e.g., intelligence processing systems), it is essential that intelligent data/information/knowledge management is an inherent component of the design from the very beginning. Ontology/Taxonomy management provides the methodologies and constructs to manage large, diverse data sets (the Big Data problem) across elements in complex System of Systems development [62].

Ontology is a discipline of Philosophy that deals with what is, the kinds and structures of objects, properties, and other aspects of reality. While much of the philosophical practice of ontology dates back to Aristotle and what his students called "metaphysics," the term *ontology* (ontologia) was coined in 1613 by Rudolf Gockel and apparently independently by Jacob Lorhard. According to the OED,[5] the first recorded use in English was in 1721. Ontologies that support Fusion types of activities include questions such as, "what is a signal?" and "what is a report?" The way we answer these perceives its environment and the way it interacts with the rest of the world.

By the early 1980s, Systems Engineering organizations dealing with knowledge representation had realized that work in ontology space was relevant to the necessary process of describing the world of Knowledge Management. This awareness and integration grew and spread System of Systems in the latter half of the final decade of the twentieth century, the term "ontology" actually became a buzzword, as enterprise modeling, e-commerce, emerging XML meta-data standards, and knowledge management, among others, reached the top of many system design requirements. In addition, an emphasis on "knowledge sharing" and interchange has made ontology an application area in its own right.

In general, the accepted industrial meaning of ontology" makes it synonymous with "conceptual model", and is nearly independent of its philosophical antecedents. We make a slight differentiation between these two terms, however (as shown in Fig. 3.5), a conceptual model is an actual implementation of an ontology that has to satisfy the engineering trade-offs and system requirements, while the design of an ontology is independent of run-time considerations, and its only goal is to specify the conceptualization of the data/information/knowledge underlying the requirements. The case for ontological analysis is strongly based on philosophical underpinnings, and a description-logic-based system that can be used to support this methodology.

3.6.1.2 Underlying Notions

We begin by introducing the most important philosophical notions: *identity, essence, unity,* and *dependence.* The notion of identity adopted here is based on intuitions about how we, as Multidisciplinary Systems Engineers, in general interact with

[5] Old English Dictionary.

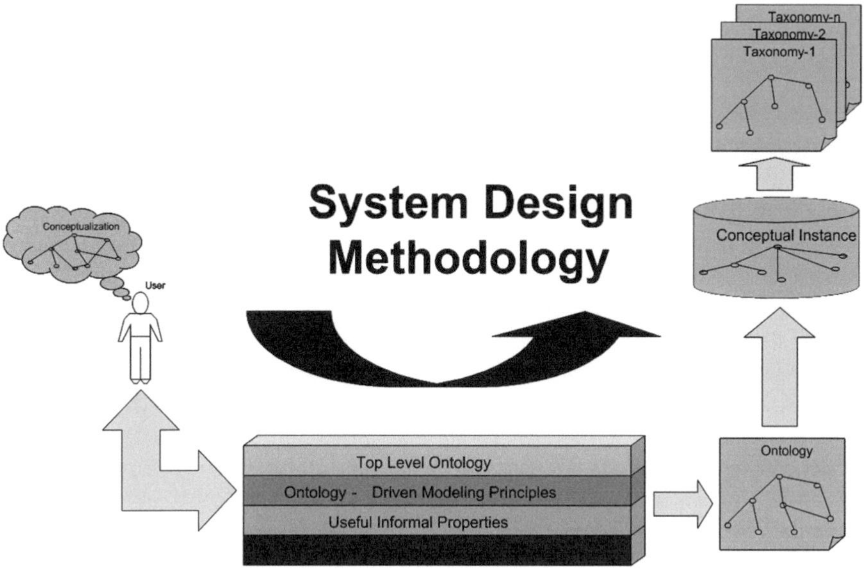

Fig. 3.5 High-level ontology development for systems design

(and in particular recognize) individual data/information/knowledge entities for the System of Systems. Despite its fundamental importance in Philosophy, the notion of identity has been slow in making its way into the practice of conceptual modeling for System of Systems, where the goals of analyzing and describing the overall system interactions are ostensibly the same. For the System of Systems types of architectures we will be discussing, the notion of identity is particularly important, because the system must recognize its environment and how to adapt to it when it changes.

The first step in understanding the intuitions behind identity requires considering the distinctions and similarities between *identity* and *unity.* These notions are different, albeit closely related and often confused under a generic notion of identity. Strictly speaking, identity is related to the problem of distinguishing a specific instance of a certain class of data/information/knowledge from other instances of this class by means of a *characteristic property*, which is unique for *it* (that *whole* instance). Unity, on the other hand, is related to the problem of distinguishing the *parts* of an instance from the rest of the data/information/knowledge by means of a *unifying relation* that binds the parts, and only the parts together. For example:

- Asking: "is this the same Mission/Business Situational Awareness information that I've seen before?" would be a problem of identity,
- Whereas Asking: "Is this data consistent with the current Mission/Business Situational Awareness," would be a problem of unity.

Both notions encounter problems when temporal aspects are involved. The classical one is that of *identity through change:* in order to account for changing environments, we need to admit that one element of the System of Systems may remain *the same* while exhibiting different properties at different times, depending on the overall emergent behavior of the System of Systems. But which properties can change, and which must not? And how can we re-identify an instance of a certain property after some time? The former issue leads to the notion of an *essential property,* on which we base the definition of *rigidity,* discussed below, while the latter is related to the distinction between *synchronic* and *diachronic* identity. An extensive analysis of these issues in the context of conceptual modeling has been made elsewhere. These issues become important as we strive to architect and field complex System of Systems. The notion of when do we determine that we are seeing a known data/information/knowledge object with a new characteristics, as oppose to a new type of data/information/knowledge, becomes extremely important and time critical.

The next notion, *ontological dependence,* may involve many different relations such as those existing between Elements and subsystems, interfaces and the range of data coming across the interface consists of, and so on. We focus here on a notion of dependence as applied to properties. We distinguish between *extrinsic* and *intrinsic* properties, according to whether they depend or not on other objects besides their own instances. An intrinsic property is typically something inherent in individual data/information/knowledge objects, not dependent on other objects, such as having temporal characteristics common to both. Extrinsic properties are not inherent, and they have a relational nature, like "where the satellite is at time *t*, or which mode the satellite transmitter is in at time *t*". Some extrinsic properties are assigned by external agents or agencies (e.g., FAA), such as having a specific location that does not change (e.g., location of an international airport).

It is important to note that our ontological assumptions related to these notions ultimately depend on our *conceptualization* of the environment in which the system will operate. It is possible the notion of "having the same location" may be considered an identity criterion for AIRPORT A, it does not mean to claim that this is a universal identity for all AIRPORTS, but that if this were to be taken as an identity criterion in come conceptualization of AIRPORTs, what would that mean for the property, for its instances, and its relationships to other object properties. These decisions are ultimately the result of our notion of the system requirements, and the expected data/information/knowledge environments. The aim of this methodology is to clarify the formal tools that can both make such assumptions explicit, and reveal the logical consequences of them.

3.6.1.3 Ontology Analysis

Here we discuss a formal ontology analysis of the basic notions presented above, and we introduce *meta-properties* that represent the behavior of a data/information/knowledge object with respect to these notions. Let us assume that we have a first-order language L_0 (the modeling language), whose intended domain is the System of Systems architecture to be modeled, and another first-order language L_1 (the meta-language) whose constant symbols are the predicates of L_0. Our meta-properties will be represented by predicate symbols of L_1. Primitive meta-properties will correspond to *axiom schemes* of L_0. When a certain axiom scheme holds in L_0 for a certain property, then the corresponding meta-property holds in L_1. This correspondence can be seen as a system of *reflection rules* between L_0 and L_1, which allow us to define a particular meta-property in our meta-language, avoiding a second-order logical definition. Meta-properties will be used as *analysis tools* to characterize the ontological nature of properties in L_0, and will always be defined with respect to a given conceptualization. We denote primitive meta-properties by bold letters preceded by the sing "+", "_", or "~". In our analysis, we adopt first-order logic with identity. This will be occasionally extended to a simple temporal logic, where all predicates are temporally indexed by means of an extra argument. Here the identity relation will be assumed as time invariant: if two things are identical, they are identical forever.

In order to avoid trivial cases in the meta-property definitions, we implicitly assume the property variables are restricted to *discriminating properties*, properties ϕ such that the discriminating properties are properties for which there is possibly something which has this property, and possibly something that does not have this property.

3.6.1.4 Knowledge Analysis

In order to capture knowledge representation, we introduce the notion of a "Knowledge Space" that represents the terminology, concepts and relationships among those concepts relevant to Knowledge Management. We illustrate Upper and Lower Ontologies to provide insight into the nature of data/information/knowledge particular to the Knowledge Management domain. Here we provide views are various levels within the overall Knowledge Management domain that the Multidisciplinary Systems Engineer might use in the overall System of Systems architecture design. The derived System of Systems archtiecture should include a Knowledge Management Upper (Fig. 3.6) and Lower Ontologies.

3.6.2 Knowledge Management Conceptual Architecture

The Knowledge Management Lower Ontology should be described in terms of different aspects the System of Systems architecture they address for solving different types of collaboration and integration issues. This is important because time after time it has been demonstrated that creating a single, monolithic System of Systems element data/information/knowledge object model and files to deliver the business/mission utility that is required is difficult, and often not infeasible. Ontologies are, by nature, much larger and more complex than data object models, and can take significant effort to build. MDSE should incorporate Component-based Ontology development approaches that are workable for each SoS development.

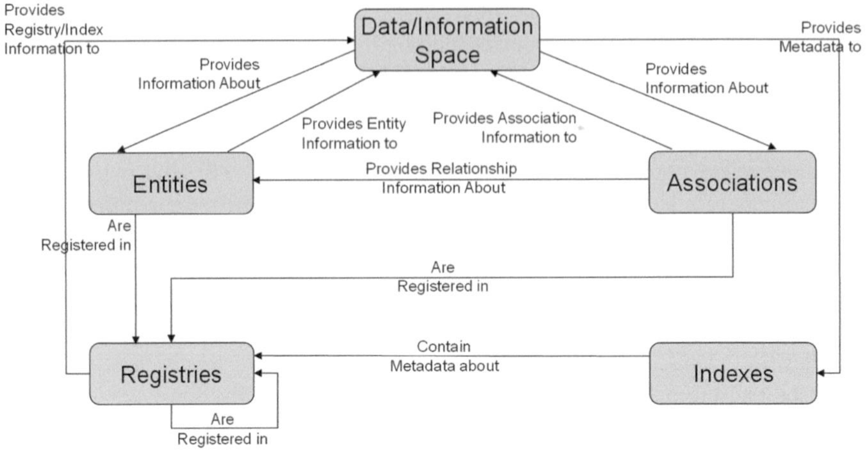

Fig. 3.6 Knowledge management upper ontology

An overall System of Systems Knowledge Management Ontology should include the following ontologies:

- **Role Based Ontology**: defines terminology and concepts relevant for a particular end-user (person or consumer application).
- **Process Ontology**: defines the inputs, outputs, constraints, relations, terms, and sequencing information relevant to a particular business process or set of processes
- **Domain Ontology**: defines the terminology and concepts relevant to a particular topic or area of interest (e.g., Satellite Communication).
- **Interface Ontology**: defines the structure and content restrictions (such as reserved words, units of measure requirements, other format-related restrictions) relevant for a particular interface (e.g., application programming interface (API), database, scripting language, content type, user view or partial view—as implemented for a portal, for instance).

3.6.3 Knowledge Management Upper Ontology

Figure 3.7 illustrates an overall Knowledge Management Ontology to manage a System of Systems for a satellite processing system. Each object within the Upper Ontology represents a major area of data/information/knowledge the system must manage. The ontologies and taxonomies illustrated below are intended to be examples, and there is not guarantee that all possible entities have been captured. Included are the entities (knowledge objects), but along with each entity are associations

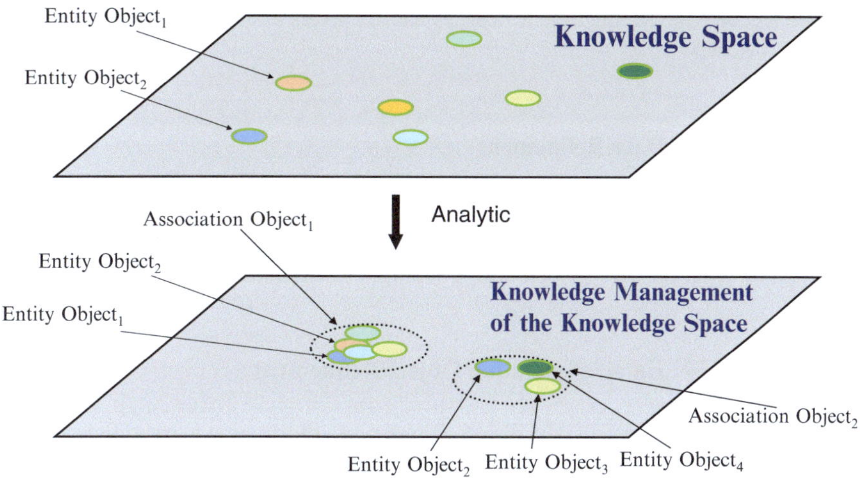

Fig. 3.7 Knowledge space management

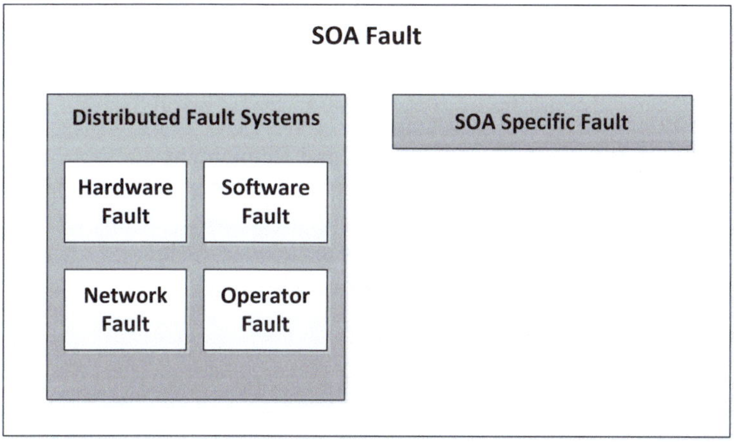

Fig. 3.8 Distributed SoS enterprise major fault categories

between entities, how the entities and associations are indexed throughout the System of Systems architecture, and the registries that tag the data/information/knowledge objects with metadata for extraction later. Figure 3.8 illustrates the relationship between the overall System of Systems knowledge space and the data/information/knowledge object associations. For each entity object, there are association objects that provide connections between objects and the metadata affiliated with those associations. Those association connections are accomplished through an analytical engine, based on the domain space of each element of the System of Systems.

The Knowledge Management Process has six levels associated with the process, Level 0–Level 5:

3.6.3.1 Level 0: Data Refinement

This entails creating a belief system where the different and varying object accuracy/belief semantics can be normalized into a data-specific semantic model that can then be used for data association, tracking, classification, etc.

3.6.3.2 Level 1: Data/Information Object Refinement

Refinement here consists of data object association, information extraction from data object associations, information object classification, indexing, and registering.

3.6.3.3 Level 2: Situational Refinement

Situational Refinement includes information object-to-object correlation and the inclusion of all relevant data into an informational display.

3.6.3.4 Level 3: Knowledge Assessment and Refinement

Knowledge Assessment consists of collecting the activities of interest, relative to each collection of information objects. This includes correlation of situational context and creation of knowledge association metadata.

3.6.3.5 Level 4: Process Refinement and Resource Management

Process Refinement enables process control and process management of data/information/knowledge movements throughout the System of Systems, inter- and intra-element communication of data/information/knowledge objects [63]. Resource allocation and management is required at this level for effective movement of data/information/knowledge throughout the System of Systems communications infrastructure and involves performance trade-offs.

3.6.3.6 Level 5: Knowledge, Decisions, and Actions

This level of processing allows information to be incorporated with system experience into system-level knowledge required to provide actionable knowledge and decision support to operators of the System of Systems.

3.6.4 *Upper Services Fault Ontology*

In order to properly identify fault messages that must be incorporated into the System of System Enterprise Service Architecture, the SoS infrastructure architecture needs an Enterprise Service Fault Ontology. The service faults apply to Service-Oriented architecture designs, independent of the technologies used (e.g., Web Services). Because there are very specific faults associated with services, the Service Fault Ontology is required to understand what classes and types of faults must be caputred and reported by the System of Systems enterprise (element-element, subsystem-subsysten, etc.). Service Level Agreements (SLAs). SoS Enterprise Service Management is a superset of SLA management and captures the information at the Enterprise Infrastructure level.

3.6.4.1 Enterprise Fault Categories

Faults may occur during any and all steps within any given SoS Enterprise service. The Service Management System must be capable of detecting service-related faults and errors. This necessitates the need to identify the fault/error categories that must be detected. The Service Infrastructure must be capable of handling these faults/errors. Besides service-specific faults, all faults that occur in the distributed SoS Enterprise may appear as service faults that will not be picked up by the normal Enterprise Management Systems. Figure 3.9 below illustrates this.

Service implementations represent a new class of problems. Dynamic linking between service providers and consumers (loose coupling) produces dynamic behavior that can cause faults in all five steps within the SoS Enterprise:

- Service Publishing
- Service Discovery
- Service Composition
- Service Binding
- Service Execution

The faults in each step can be caused by a variety of reasons. Table 3.1 illustrates the percentages of fault reasons [64].

The SoS Enterprise Service Fault Upper Ontology starts with the five general steps within the SoS Enterprise and then each category is refined. This generalization allows a complete coverage of possible service faults. Each service within the SoS will have its own SLA with its own performance objectives, called Service Level Objectives (SLOs). This provides domain-specific faults relative to a particular domain with the SoS architecture. Figure 3.9 below illustrates an example of a Service Fault Ontology.

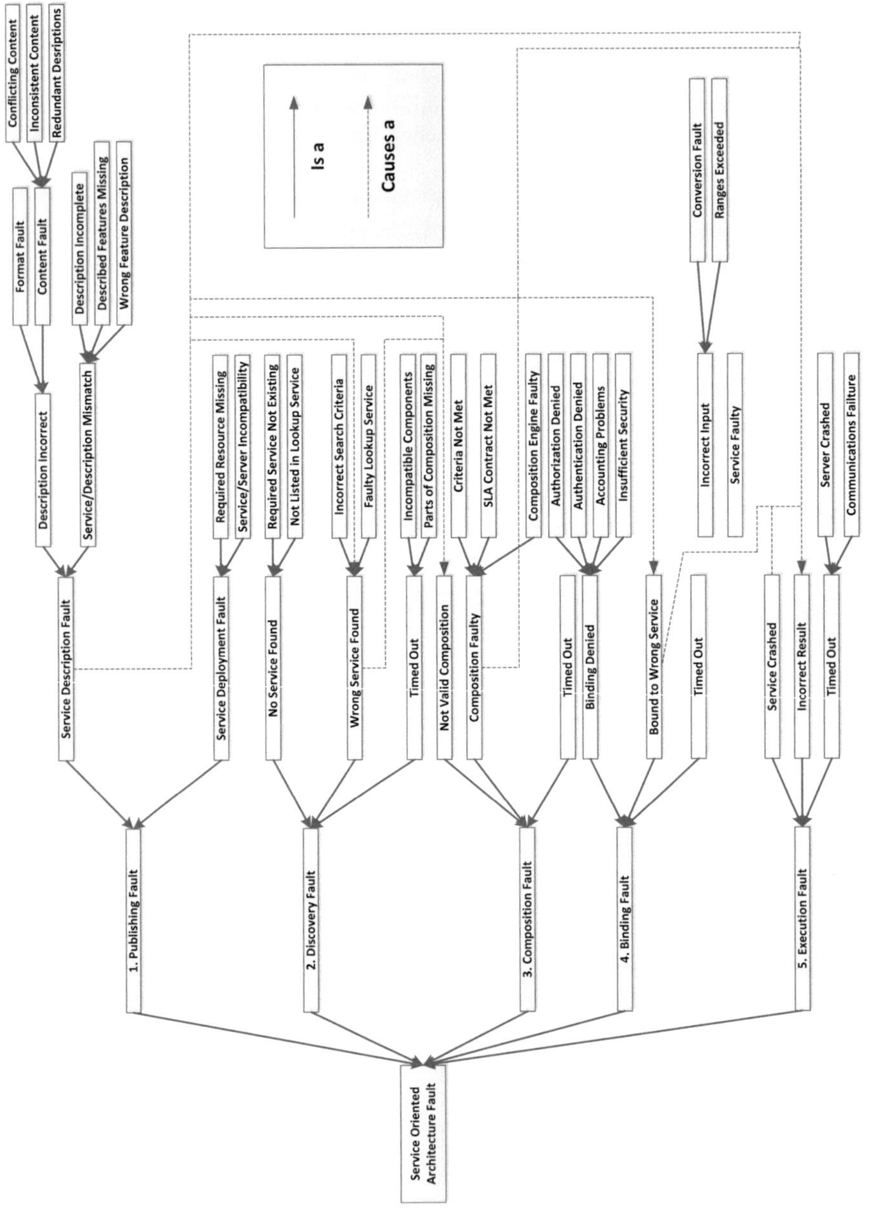

Fig. 3.9 SoS enterprise service fault ontology

3.7 Discussion

Embracing the MDSE philosophy is challenging, as it requires the MDSE teams to take multiple roles simultaneously. This requires changes in team structures, management structures, electronic productivity tools, and new methodologies for systems, software, integration and test, operations and maintenance; every aspect of the System of Systems life cycle. Commercial companies are ever increasing their awareness and desire for Systems Engineering skills within their organizations. As this trend continues, the diversity of applications for Systems Engineering will drive the necessity for an MDSE approach. The MDSE roles and responsibilities are many, but they provide a multidisciplinary view of the overall System of System life cycle and minimize the risk of developing products that are not operationally viable by assuring timely involvement of all disciplines in the requirements discovery/decomposition/derivation phase of the design process. Complex designs require complex, multidisciplinary thinking and complex roles; something MDSE embraces.

Chapter 4
Systems Engineering Tools and Practices

In order to deal with the complexity of systems engineering designs for modern systems, formalized methods and tools have been created to allow documentation, communication, and comparison of systems and software architectures [65]. These methods and tools all focus around the creation of different drawings, each with their own set of standards, each designed to aid in the design, implementation, and manufacturing phases of overall system development, operations, and maintenance. As has been discussed throughout this textbook, the system must be defined through with a complete multidisciplinary lifecycle perspective in order to capture all of the requirements, needs, and goals of the customer. The overall objective is to develop a complete lifecycle-balanced system. As system complexities have increased over the decades, the Systems and Software Engineering tools have evolved to keep up with changing system designs (e.g., service-based systems, agile development methods, etc.) [35].

Process standards have been developed which aid the multidisciplinary systems engineer in establishing baselines for system design. One example is the IEEE 29148-2011—Systems and Software Engineering—Lifecycle Processes—Requirements Engineering standards. Others include the IEEE 12207—Systems and Software Engineering—Software Lifecycle Processes, which is an international standard for establishing a common framework for software lifecycle processes. The EIA 632 Processes for System Engineering was established by the Engineering Industries Alliance (EIA) in conjunction with the International Council on Systems Engineering (INCOSE) to focus on standards for conceptualization and implementation of complex systems engineering designs [66].

Large amounts of design data are created during the lifecycle of any complex system. Different engineering methodologies are utilized during each lifecycle phase, each with their own documentation tools and defined "views" of the overall system. No one tool captures the entire systems, software, operations, and maintenance views of the system. It is incumbent on the Multidisciplinary Systems Engineer to manage all the design and implementation documentation to provide complete traceability; horizontally and vertically through any stage in the lifecycle

J.A. Crowder et al., *Multidisciplinary Systems Engineering*,
DOI 10.1007/978-3-319-22398-8_4

process. This includes not just initial design views, but systems integration, verification and validation plans, operations and maintenance specifications, and overall human-machine interface plans. Management of all of these different views and specifications is vital for the engineering of the overall system. Such processes and methodologies are not linear, requiring multiple iterations between each set of specifications in order to ensure a viable, usable, working system. The multidisciplinary approach is necessary in order to coordinate and facilitate the entire system traceability across the entire product lifecycle.

Each set of design and implementation documentation requires computer-based tools in order to produce the required design "views," each based on a different set of standards, whether U.S. Department of Defense (DoD), U.S. Commercial standards, or International standards. There are a large number of systems and software design tools available in the market. Each is based on a given methodology and focus (e.g., systems vs. software engineering). The following sections will describe the main methodologies for each of the computer-aided productivity tools are based upon. One of the important points to understand is that no one tool or methodology will give a complete view or complete documentation of a complex systems and software engineering design. It should be expected that development information for any complex system will be distributed over multiple tools, covering multiple engineering disciplines; hence the need for the multidisciplinary approach. The first methodology to be discussed will be the U.S. Department of Defense Architectural Framework (DoDAF).

Question to think about:
Does having formal documentation help or hurt a system development?

4.1 System Architecture Frameworks

In order for the Systems Architect/Analyst to create the architectures required for system development (systems, software, information, data, etc.), there must be an established framework (or structure) that can be used to communicate the architecture to the rest of the program and to the customer. An architecture framework defines how to create a given architecture, providing the principles, practices and standards for creating, using, and communicating an architectural description of an engineered system. It allows the architect to structure their thinking and architecture vision by allowing the Systems Architect/Analyst to divide up the architecture into separate layers or views, and provides standards methodologies for creating and documenting the architecture in terms of diagrams, matrices, and word documents. A high-level view of an architectural layer view is shown in Fig. 4.1.

There are several different Systems Architecture Frameworks that are utilized across commercial and military systems in the United States and in other countries. These include, but are not limited to, the Zachman Framework, the Department of

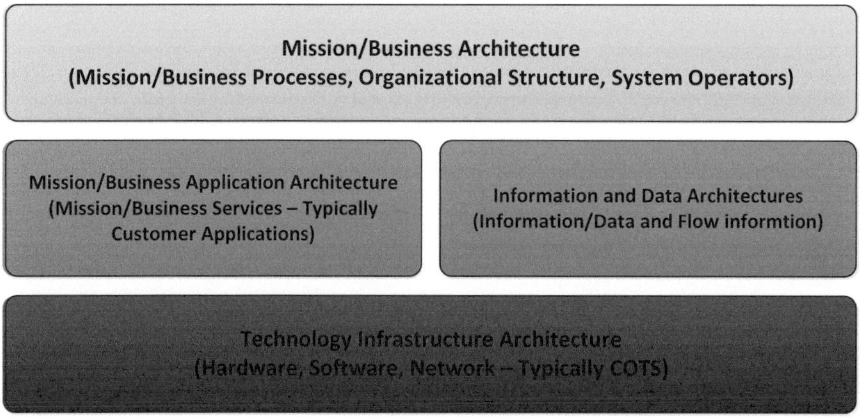

Fig. 4.1 High-level systems architecture layer view

Defense Architecture Framework (DoDAF), The Open Group Architecture Framework (TOGAF), and the Ministry of Defense Architecture Framework (MODAF). A brief description of each is provided below [46].

4.1.1 The Zachman Framework

The Zachman Framework represents a classification methodology for system architecture descriptions [67]. The Zachman Framework matrix, shown in Table 4.1, is constructed to classify the architecture documentation by the audience the documentation is intended for, and the content focus of the artifact. The Zachman Framework does not specify which of the architecture frameworks (e.g., DoDAF, TOGAT, MODAF, etc.) to use for artifact creation, it just lays out a detailed analysis of which artifacts are applicable to which perspective and which models (audience) [68]. For instance row 2, the Conceptual Models can be thought of as the Owner or Customer's perspective since these artifacts relate to the conceptual design and capabilities for the design of the system. The customer and development team should come to concurrence at this level, in order to ensure that the system being developed is in line with the customer's Concept of Operation (CONOPS), technical and performance needs, as well as usability by the end users. Understand, that by customer, we mean who ever has commissioned, or is paying for the system. This might be an outside customer like NASA, the Department of Defense, NATO, etc., or it might be an internal customer where upper management is paying for the system (e.g., Information Technology computer and network upgrades) [69].

Row 1 defines the scope of each perspective, where the information contained in each perspective will drive the conceptual views and will drive the requirements specified for the system. The artifacts specified for each perspective in row 3, called the Designers Perspective, lay out the overall architectural views and designs that

Table 4.1 High-level Zachman framework matrix

	Data perspective	Functional perspective	Network perspective	Organizational perspective	Temporal perspective	Motivational perspective
Contextual model	Mission/business data/information	Mission/business processes	Mission/business physical locations	Mission/business organizational structure	Important business/ mission events	Mission/business strategies
Conceptual model	Conceptual information/ data model	Mission/business process model	Mission/business logistics model	Mission/business workflow/ orchestration model	Mission/business master schedule	Mission/business functional model
Logical model	Logical data models	System architecture model	Distributed network/systems model	Human-systems interface model	Information/data processing architecture	Mission/business rules
Technology model	Physical information/ data model	Technology model (plans)	Technology infrastructure model	Presentation/ visualization architecture	Process control model/architecture	Processing rule strategy
Representational model	Data item descriptions	Program execution model	Network topology	Security architecture	Timing diagrams	Processing rule specification
Functional model	Data instantiation	Functional system architecture	Functional network architecture	Functional organization	Functional schedule	Functional/working business/mission strategies

take the concepts in row 2, translates them into a design, allowing the perspectives in row 4 to define what is technically and physically feasible for the project. Row 5 disassembles the designs into physical parts (hardware and networks), and software components (services), that are required to instantiate the overall design; that is, how to physically put the system together. Row 6 represents the final instantiation of row 2. Row 6 is the demonstration to the customer that the system meets their needs, as specified in the requirements and CONOPS.

4.1.2 DoDAF

The Department of Defense Architecture Framework (DoDAF) provides an Architecture Framework standard for defense contractors to organize and document system architectures for proposals, acquisitions, and system design/development. It allows comparisons of different defense contractors' designs without having to translate between different architectural frames or reference. The DoDAF 2.2 Framework specifies artifacts and documentation that provide views from various perspectives for the system (e.g., Data Views, Capability Views, Service Views, etc.). These are described at a high-level in Fig. 4.2.

The DoDAF 2.2 Framework contains the following architectural views and documentation shown in Table 4.2 below.

The DoD Architecture Framework (DoDAF) is the US Department of Defense's customized framework for developing integrated system architectures. DoDAF describes the rules and conventions that defense and defense contractor systems

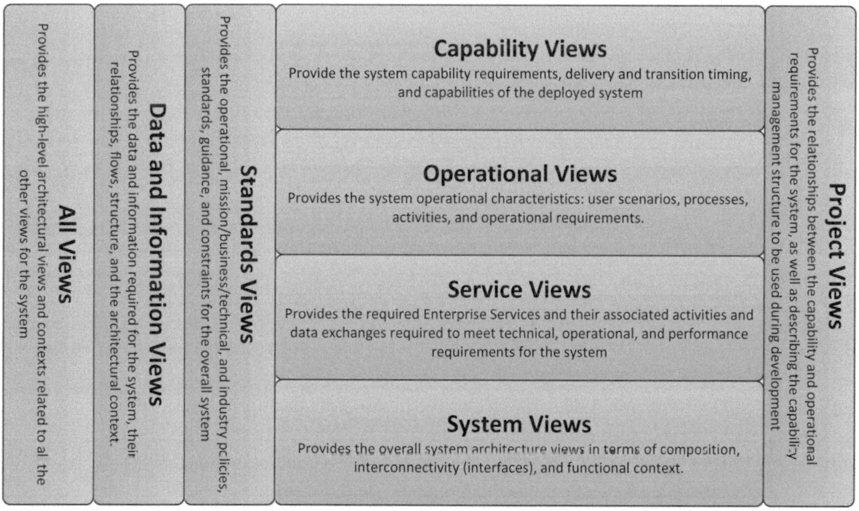

Fig. 4.2 DoDAF 2.2 architectural framework documentation views

Table 4.2 DoDAF architectural documentation views

Views	Description
All views	
AV-1	Describes the project goals, objectives, activities and events, plans, constraints, measures of effectiveness, and expected outcomes (products)
AV-2	Repository that includes all terms with definitions used throughout the program
Data/information views	
DIV-1	Conceptual data model(s)
DIV-2	Logical data model(s)
DIV-3	Physical data model(s)
Standard views	
Stdv-1	List of standards that apply across the program
Stdv-2	List of possible emerging standards and their potential impacts on the program
Capability views	
CV-1	Provides the strategic context for capability development for the program. Should provide the context for the capabilities shown in the architecture views
CV-2	Models a hierarchy of capabilities (the capabilities taxonomy)
CV-3	Shows time phasing of capabilities development across the program, including the activities, conditions, desired effects, and necessary resources
CV-4	Describes the dependencies between the planned capabilities. Also provides the definitions for logical groupings of capabilities
CV-5	Planned capability deployment. Shows the planned capability phasing in terms of personnel and locations and their concept of capability development
CV-6	Provides a mapping between the capabilities and the operational activities throughout the program
CV-7	Provides a mapping between the capabilities and proposed service development across the development efforts
Operational views	
OV-1	Operational concepts. Graphical and textual description of the overall program operational concepts
OV-2	Operational resource flows. Description of the resources exchanged between operational activities
OV-3	Operational resource attributes. Provides a description of the resource attributes relevant to the resources exchanged in the OV-2
OV-4	Describes the relationships between different organizations being utilized across the program (e.g., subcontractors)
OV-5a	Operational activity decomposition. Hierarchical view of the proposed operational activities for the systems
OV-5b	Operational activity model. Describes the proposed system operational capabilities and their relationships to inputs, outputs, and system activities
OV-6a	Operational rules model. Describes the mission/business rules that constrain the overall system activities
OV-6b	State transition model. Describes the mission/business processes and their responses to events and/or activities experienced in system operations
OV-6c	Event trace model. Describes the traces between operational activities and sequences of events across the proposed system operations

(continued)

Table 4.2 (continued)

Views	Description
Service views	
Svcv-1	Service context. Identifies the proposed system services and their interconnectivity/ interdependencies
Svcv-2	Service resource flow. Describes how resources flow between proposed system services
Svcv-3	System service matrix. Describes the relationships between the system, subsystems, and components, and the proposed system services for each architectural view
Svcv-4	Service functionality description. Describes the functionality of each proposed system services and the associated data flows
Svcv-5	Operational activity to services matrix. Maps the proposed services to proposed system operational activities
Svcv-6	Services resource flow matrix. Describes the resources and the resource flow between proposed services and the attributes of those resource exchanges
Svcv-7	Services measures matrix. Describes the measures of effectiveness for proposed services and the timing of the measures
Svcv-8	Services evolution. Describes. Describes the incremental development of proposed services. Includes a description of the evolution of services towards more efficient overall system operations in the future
Svcv-9	Services technology and skills forecast. Describes the emerging technologies possible for use on the program, including hardware, software, personnel skills and the associated time frames for each. This is specific to how it affects and/or improves services throughout the system
Svcv-10a	Service rules model. Identifies constraints that are imposed on the proposed system service functionality, caused by the proposed system design and/or implementation
Svcv-10b	Service state transitions. Describes the service state transitions caused by service responses to system events
Svcv-10c	Service event-traces. Describes service-specific sequence of events driven by the operational views
System views	
SV-1	System interfaces. Describes to systems, subsystems, and system items and their proposed interactions
SV-2	System resource flows. Describes the resource flows between system elements
SV-3	System-system matrix. Describes the relationships between systems, elements, subsystems, etc. given in the architecture views. This may include planned vs. existing interfaces, both internal and external
SV-4	System functionalities. Describes the functionality of the systems, subsystems, etc. and the data flows between proposed system activities
SV-5a	Operational activity to functionality matrix. Describes the mapping between the proposed operational capabilities and operational activities
SV-5b	Operational activity to system tracing. Describes the mapping of systems, subsystems, etc. to operational activities
SV-6	System resource flows matrix. Describes resource exchanges between the systems and the attributes of those exchanges
SV 7	System measures. Describes the metrics that will be utilized for the system elements and their timeframe(s)

(continued)

Table 4.2 (continued)

Views	Description
SV-8	System evolution. Describes the program's plans for migrating from the current design to a more efficient system, or towards evolving a current system to a future, better implementation
SV-9	System technology and skills assessment. Describes emerging technologies, such as new hardware and software (usually COTS) that are expected to be available in the future (with time frame estimates) and how that will affect the overall system activities and performance
SV-10a	System rules model(s). Describes constraints that are imposed on the system and it's overall design. This are in terms of system functionality
SV-10b	System state transition. Describes the service affects and transitions in response to expected events
SV-10c	System event trace. Describes system-specific critical sequences-of-events that are documented in the operational views of the system
Project views	
PV-1	Describes the relationships between the program organizations and outlines the program organizational structures that are required to manage the program and maps these to program capabilities
PV-2	Program timelines. Describes key milestones, their timing throughout the program, and their interdependencies
PV-3	Program capability mapping. Describes the mapping between program capabilities to illustrate how program elements and/or specific projects within the program are utilized to achieve specific capabilities required by the system

architects use when developing architectures for DoD systems. DoDAF was created so that the resulting work is compatible, at least at a basic level. DoDAF was created to support many facets of the DoD, and was intended to be useful even when limited parts of the architecture framework are implemented. The current instantiation of DoDAF, DoDAF 2.2 is based on a six-step architecture design process. The DoDAF 2.2 guidelines differs from previous versions of DoDAF in that 2.2 emphasizes relationships between data rather than products (data-based vs. view-based). This emphasis on a data-centric architecture helps to ensure consistency between the Systems, Software, and Information architectures, as well as providing proper data/information flows in/out of both external and internal interfaces. Figure 4.4 illustrates the DoDAF 2.2 Systems Architecture development process. Following the development process outlined in Fig. 4.3 helps to define the views that are appropriate for the system to be designed. Only those views that are required to represent the applicable data/information in the system should be baselined for the architectural design. The architecture views represent the architecture for a given time-period, realizing that the needs of the system, the requirements, and the operational concepts may change over time and the architecture must evolve as these characteristics change. Agile Systems Engineering methodologies and techniques should be built into the overall process in order to accommodate changes to the system [56].

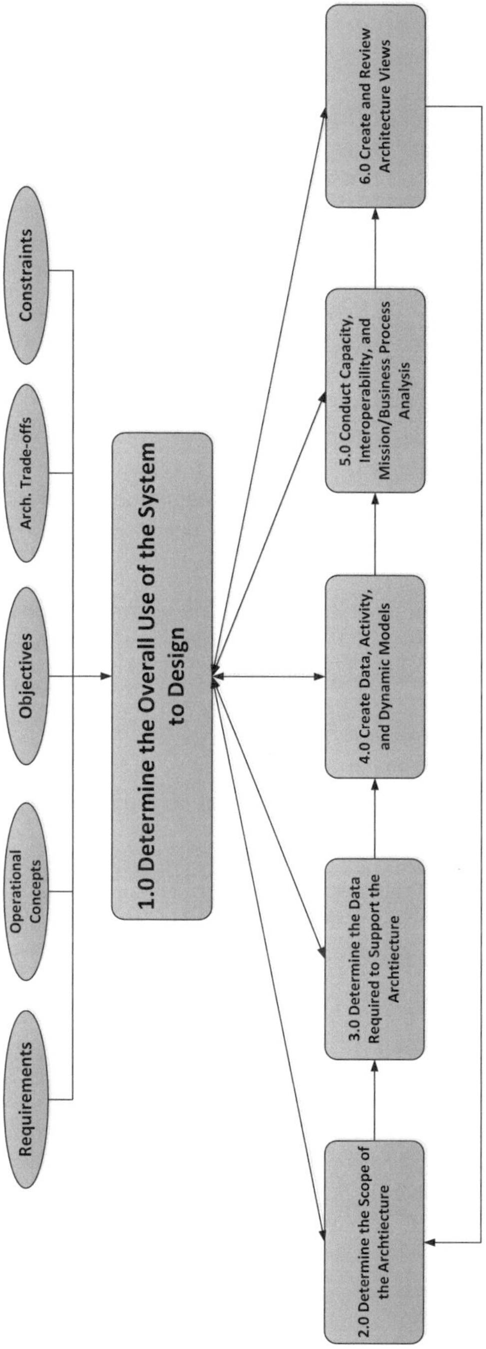

Fig. 4.3 Simplified view of the DoDAF architecture development process

4.1.3 TOGAF

DoDAF concentrates on providing architecture documentation via various views, as shown in Table 4.1; it does not specify the methodology for producing these views. In contrast, TOGAF focuses on the methodologies available to provide architecture documentation, without specifying which views, or constructs, should be used to design system architectures. TOGAF evolved out of the Department of Defense Technical Infrastructure Framework for Information Management (TAFIM). Version 3.0 of TAFIM came out in 1996, was base-lined. turned over to TOGAF and retired. It is utilized by the TOGAF group to develop their TOGAF Architecture Development Method (ADM). The purpose of the ADM is to provide a guide to Systems Architects to lead them through the system architecture lifecycle. The TOGAF ADM provides a mapping between their methodologies and the DoDAF Architecture Views. Figure 4.4 illustrates this mapping.

TOGAF is a high level approach to design. It is typically modeled at four levels: Business, Application, Data, and Technology. It relies heavily on modularization, standardization, and already existing, proven technologies and products. It is a set of tools which can be used for developing a broad range of different architectures. It should:

- describe a method for defining an information system in terms of a set of building blocks
- show how the building blocks fit together
- contain a set of tools
- provide a common vocabulary [70]
- include a list of recommended standards
- include a list of compliant products that can be used to implement the building blocks

4.1.4 The Ministry of Defense Architecture Framework (MODAF)

MODAF is an international architecture description standard that has been utilized within the international community to support the delivery of military capabilities across the world and has been adopted by NATO to form the core of their NATO Architectural Framework (NAF). Currently the United States Department of Defense, the Canadian Department of National Defense, the Australian Department of Defense, and the Swedish Armed forces are working together to utilize MODAF and DoDAF to form the International Defense Architecture Specification (IDEAS) which will converge TOGAF, MODAF, and DoDAF into a single, unified architecture framework. Figure 4.5 illustrates the basic MODAF views.

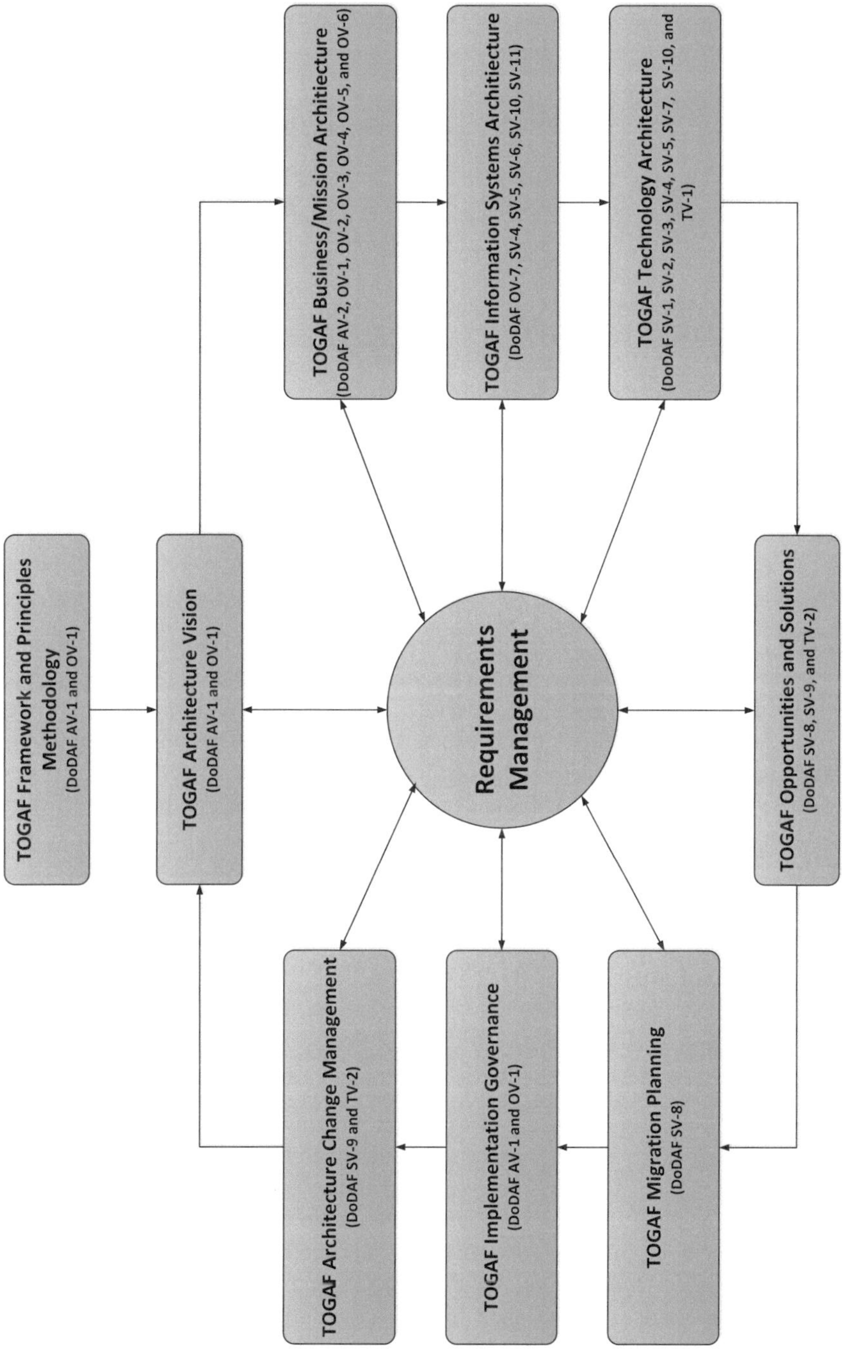

Fig. 4.4 TOGAF to DODAF mapping

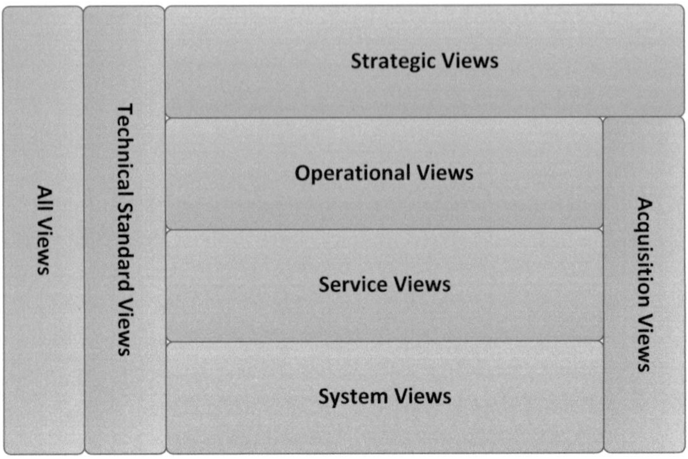

Fig. 4.5 Overview of the MODAF views

4.1.5 The International Defense Enterprise Architecture Specification (IDEAS)

IDEAS is a multinational organization aimed at forming a unified System Architecture Framework and data exchange format for military and civilian system architecture design use. Creation of a unified architecture framework will allow seamless sharing of data, architectures, and designs between partner nations. The initial scope of the IDEAS group is to provide translational capabilities between architecture frameworks (e.g., DoDAF and TOGAF), and the initial charter of the group is for exchange of systems architectural data in order to support multi-nation coalition planning and operations. Long-term plans are to look at merging the DoDAF 2.2 meta-model and the MODAF Ontological Data Exchange Mechanism (MODEM) into a unified architectural modeling tool (similar to how UML unified software architecture) [71].

4.1.6 UML

The **Unified Modeling Language** (**UML**) is a general-purpose modeling language in the field of software engineering, which is designed to provide a standard way to visualize the design of a system. It was created and developed by Grady Booch, Ivar Jacobson and James Rumbaugh at Rational Software during 1994–1995 with further development led by them through 1996 [72]. In 1997 it was adopted as a standard by the Object Management Group (OMG), and has been managed by this organization ever since. In 2000 the Unified Modeling Language was also accepted by the International Organization for Standardization (ISO) as an approved ISO standard. Since then it has been periodically revised to cover the latest revision of UML [73]. Much more will be presented on UML in Chap. 5.

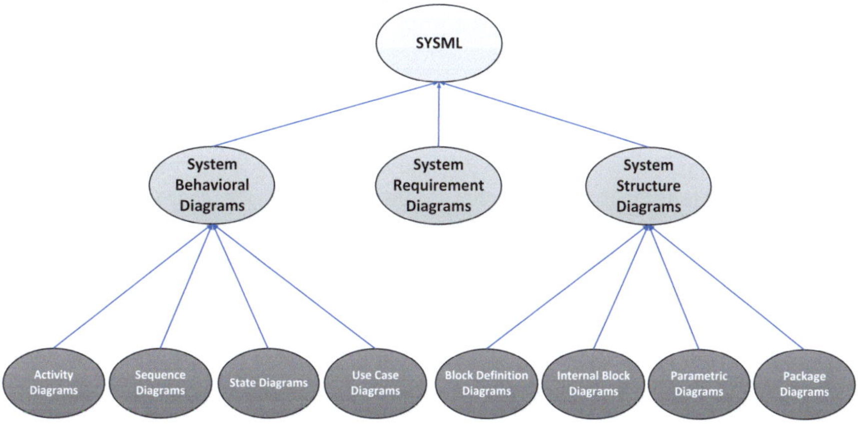

Fig. 4.6 9 SYSML views

4.1.7 SYSML

SysML is based on UML and involves modelling blocks instead of modelling classes, thus providing a vocabulary that's more suitable for Systems Engineering. A block encompasses software, hardware, data, processes, personnel, and facilities. SysML reuses a subset of UML2 (UML4SysML), and defines its own extensions. Therefore, SysML includes 9 diagrams instead of the 13 diagrams from UML2, making it a smaller language that is easier to learn and apply. SysML can be easily understood by the software community, due to its relation with UML2, whilst it is accessible to other communities, SysML makes it possible to generate specifications in a single language for heterogeneous teams, dealing with the realization of the system hardware and software blocks (see Fig. 4.6). Knowledge is thereby captured through models stored in a single repository, enhancing communication throughout all teams. In the long term, blocks can be reused as their specifications and models enable suitability assessment for subsequent projects [74].

Questions to think about:
Does the documentation need to be maintained in relation to the approved changes on the program?

What happens when the documentation is allowed to decay (become outdated)?

What happens to the system Life Cycle Cost (LCC) model when the decay has occurred in the documentation?

4.2 System Architecture Analysis Methodologies and Productivity Tools

The development of System of Systems architecture is a relatively new field in systems engineering that has come about in the last two decades in relation to System Architecture. Many methodologies have come and gone over that time, where the discussions above, on the major remaining architecture analysis and design methodologies, are being applied to System of Systems developments [75]. One thing to note about the above discussion is that for many large, complex system designs, none of the available design methodologies provides a complete structure for a given system of systems design. There are multiple text and professional books written on each of these methodologies and of the analysis methods discussed here.

4.3 Case Study: Failures in Change Management

One of the most important aspects of choosing an architecture framework for a given System of Systems design is ensuring that the system architecture documentation keeps up with changes in the System of Systems design. This involves an effective Change Management System. Without it, the MDSE can be in store for spectacular failures, as this case study will show.

4.3.1 Satellite Launch Systems

Early 1990's satellite launch: With the ability to launch multiple satellites from a single launch vehicle, this particular launch had only one spacecraft on board. In order to provide a launch platform for multiple satellites, the launch vehicle was wired with a harness that could accommodate multiple spacecraft, with the bottom spacecraft attached to the lower harness connector, and continued to the top spacecraft connected to the top part of the wiring harness, where the top spacecraft would be the first to separate from the vehicle [76].

This particular launch had only one spacecraft to be carried to space with the launch vehicle. Given there was only one spacecraft, any of the harness connectors could be utilized, as long as the hardware and flight software agreed on which harness connector was to be utilized. Initially in the design, the top wiring harness connector was to be used to connect the spacecraft to the launch vehicle and the flight software would issue the command to separate from the top connector of the wiring harness.

Somewhere in the implementation process, the hardware group decided to utilize the bottom connector of the wiring harness, because it was a better fit for the configuration of the satellite to be launched (i.e., it reached easier). Unfortunately, the Configuration Management change to the wiring harness connectivity was never put under Configuration Management control; therefore the Change Management notice was never generated to tell the software group that the different wiring harness connector was to be used for the launch. Adding to this problem was the fact that to save money, the Launch Vehicle-to-Satellite interface was not tested, since it was assumed to be correct. The result was that the launch vehicle's second stage failed to separate from the spacecraft (satellite).

The only way to separate the satellite from the launch vehicle was to release the satellite from its apogee motor that would have taken the satellite up to its geosynchronous orbit. Instead, the smaller perigee motor was used to put the satellite into a useless low-earth orbit. Later that same year, a new perigee motor was attached to the satellite (Space Shuttle crew) and it was able to then take itself to its geosynchronous orbit. All in all, this was an approximately $300,000,000.00 mistake, all for the lack of a $0.25 change notice. Multidisciplinary Systems Engineers ***MUST*** keep track of details, for small mistakes can have major effects.

4.4 Discussion

Deciding on the framework to use (e.g., TOGAF) for a given System of Systems design requires careful consideration. Just because a particular framework has many architecture views that could be used doesn't mean a given system design requires them. There are many groups that espouse one architecture framework over another. Remember each has advantages and disadvantages and careful investigation is required. If you ask someone who belongs to a particular framework "discipleship" you will only hear why their framework is the only choice. The reality is, it depends on what you are comfortable with, for all can be used; it just depends on where you want the pain of getting around their deficiencies. The bottom line is:

- Zachman—described as a framework, more accurately described as a taxonomy.
- DoDAF—described as a framework, more accurately described as an organized collection of views.
- TOGAF—described as a framework, more accurately defined as a process.
- MODAF—described as a framework, more accurately defined as a methodology.

Chapter 5
The Overall Systems Engineering Design

5.1 Designing for Requirements

The mission/business analysis community typically attempted to keep analysis distinct from technical design. It is not inherently different or an incorrect term for development of a solution to a system design. In the end, the activities which we would call design are nothing different from the activities required to create the "To-be" requirements.

Figure 5.1 illustrates this process. Figure 5.1 describes the process of analyzing an enhancement to an existing system or process. In the traditional mission/business analysis approach, the "As-Is" state is analyzed as a starting point to capture processes, data, and interactions that must be kept for any new solution. Those are shown as the "as-is" requirements in Fig. 5.1.

The requirements/functionality of the legacy system is identified, along with the processes, data, system interfaces, and applicable displays. The design process comes into play when determining how to integrate them with the new features being added. These new features become requirements/functionality that must be added to design the overall new/enhanced system. This is not technical or physical design, but a logical design discussed earlier. Examples of new requirements/functionality include many traditional outputs including:

- Process models showing the legacy processes to be re-used as well as the required new processes [77].
- Data models showing final new/enhanced system data requirements and mission/business rules relating to the relationships between entities.
- Use Case models with interaction requirements and corresponding mission/business rules.
- Prototypes and mock-ups indicating interface requirements, including presentation mock-ups.
- Interfaces with other systems or parts of the system (e.g., subsystem interfaces).

J.A. Crowder et al., *Multidisciplinary Systems Engineering*,
DOI 10.1007/978-3-319-22398-8_5

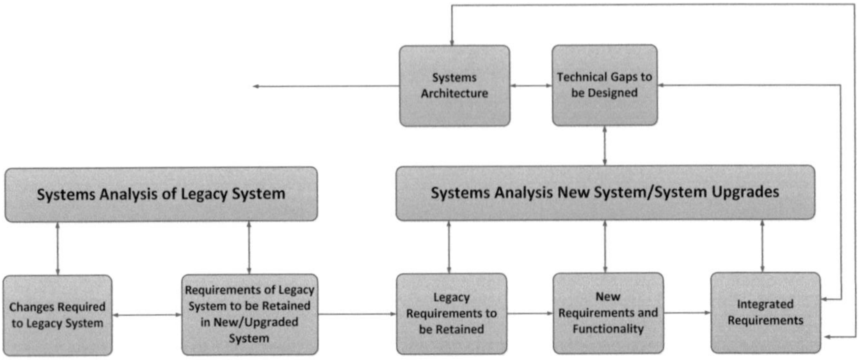

Fig. 5.1 Analytical process for reuse of legacy systems in a new or enhanced system

5.1.1 Requirements Decomposition

Decomposition of the customer's requirements is the first step toward defining the architecture after the systems analysis has been completed. The requirements decomposition process is not set in stone and is influenced by the type of system being designed, the domain in which the system must operate, the system constraints, and other factors that may be specific to any individual system. The requirements decomposition process radically affects the optimization of the architecture, system design, and system implementation, since different decompositions drive different internal interfaces, different information and data flows, and a different set of Enterprise Services to execute the functionality laid out in the systems analysis and subsequent decomposition. Each of the different decomposition possibilities drives a trio of different time, quality and cost profile for the development of the system. These must be balanced against the customer technical, performance, and quality requirements, customer-driven schedule, and available cost profile. Some of the factors that affect the balance of these three factors is the desire for system/subsystem/component/service reuse, different organizational structures required to execute (develop) the system, software development style (e.g., spiral, waterfall, agile, etc.), and technology base (i.e., COTS vs. Customer Built). These aren't the only considerations, but they are the major ones that will drive the development optimization. Figure 5.2 illustrates this idea [78].

In order to balance decomposition of the requirements into optimized system functions at each level of decomposition (e.g., system, subsystem or service group, etc.) the decomposition criteria can be broken up into four main attribute groups: customer directives, functional directives, quality directives, and technical directives, which are illustrated in Fig. 5.3

If you examine the directives shown in Fig. 5.3, you will see directives that come from many engineering disciplines, including Cognitive Systems Engineering,

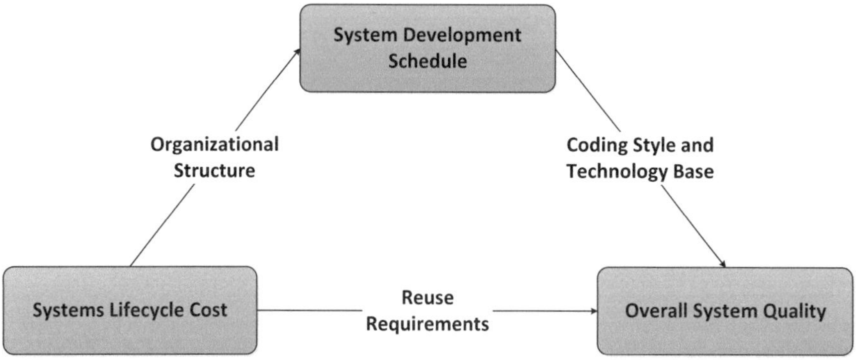

Fig. 5.2 Systems optimization trade-off analysis parameters

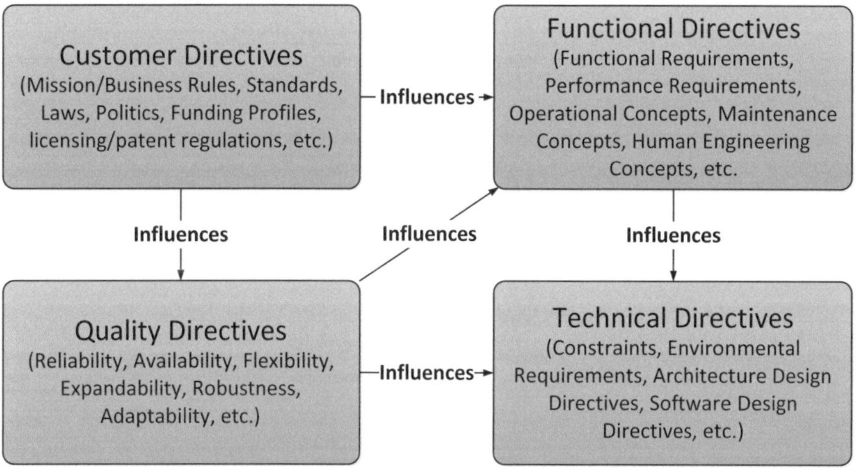

Fig. 5.3 High-level requirements decomposition criteria

Software Engineering, Safety Engineering, Sustainment Engineering, and many others. Each of these must be taken into account in the initial requirements decomposition to ensure that all of these disciplines are embedded into the initial functional block diagram and requirements derivation/allocation. As the requirements are decomposed utilizing the directives (Fig. 5.3) each decomposition level must be assessed for its viability. Given the overall requirements, environment, constraints, laws, etc. that must be taken into account, there are multiple, possibly dozens, of ways to adequately decompose a system. In order to create an objective methodology for the requirements/system decomposition, as set of weightings are created, within each directive category, agreed to by all disciplines, before the decomposition is begun. Table 5.1 below is an example of such a directive-weighting table.

Table 5.1 Requirement directive decomposition weighting example

Directive	Weighting
Functional directives	
Clustering of functions by usage	0.2
Requirements dependencies	0.3
Internal interfaces	0.1
Operational concepts	0.1
Human-machine interface requirements	0.3
Total	1.0
Technical directives	
Technical requirements/constraints	0.5
Design directives	0.3
Communication requirements	0.2
Total	1.0
Customer directives	
Laws and standards	0.1
Licenses, certificates, patents, etc.	0.1
Mission/business rules	0.1
Information requirements	0.1
External interfaces	0.2
Subcontractor relationships	0.1
Political considerations	0.1
Cost profiles	0.1
Schedule profiles	0.1
Total	1.0
Quality directives	
Performance requirements	0.3
Reliability requirements	0.2
Availability requirements	0.1
Maintainability requirements	0.1
Usability requirements	0.1
Security requirements	0.2
Total	1.0

As requirements are decomposed and derived, the following set of guidelines should be utilized to assess the adequacy of the derived requirements:

- Correctness
- Consistency
- Traceability
- Disambiguation
- Testability
- Atomicity

5.1.1.1 Completeness

For requirements to be considered complete, we define three characteristics. The first two properties, shown below, can be thought of as "*internal completeness*" and imply a closure, or completeness of available information. The last characteristic involves "*external completeness*" and attempts to ensure that all of the problem definition information is contained in the requirements:

- There is no information that is unstated, or "*To-be-Determine*" (TBD).
- The requirement text does not contain undefined entities or objects.
- No information is missing from the requirement.

5.1.1.2 Correctness

Correctness ties completeness, consistency, and traceability together to understand if the derived requirement, as part of the requirements hierarchy, not only reflects the parent requirement from which it is derived, but also traces correctly through the hierarchy, is consistent with other requirements derived from the same parent requirement, and is complete in its structure and language. There should be no missing components consistent with the functionality of the functional block each requirement is mapped into.

5.1.1.3 Consistency

Consistency measures whether any functional or non-functional requirements are in conflict with each other [79]. This applies to functional interfaces (both internal and external), as well as performance and quality attributes. If one derived requirement states that two inputs should be compared and a procedure initiated if $A>B$, and another requirement states the procedure should be initiated if $B>A$, these two requirements are not consistent. Consistency of requirements is critical to ensuring a usable system is developed.

5.1.1.4 Traceability

Ensuring proper requirement traceability is essential to integrating and testing the system, throughout the system development. Traceability is concerned with documenting the parent/child relationships between requirements at each level of the architecture. It also documents the relationships between the requirements and other architectural and design artifacts [80]. The purposes of traceability are threefold:

- Manage engineering changes across the system development
- Understand the decomposition of the system into each architecture tier
- Manage the overall quality of the developed system

Requirements traceability allows understanding of the objectives, goals, aims, expectations, and needs of the system under development at every tier in the architecture. It allows understanding of the relationship between the architecture tiers [81].

5.1.1.5 Disambiguation

An unambiguous requirement is one that is stated in concise and not esoteric language, without technical jargon, without acronyms that have not been previously defined. An unambiguous requirement contains facts, and is written without negative language or compound statements. The disambiguated requirement does not contain opinions and is not subject to interpretation.

5.1.1.6 Testability

For a requirement to be testable, the correct implementation of the requirement must be able to be determined through standard methods:

- Demonstration
- Test—includes instrumentation to validate the tests
- Inspection
- Analysis—this may include valid modeling and simulation techniques

5.1.1.7 Atomicity

A requirement is said to have atomicity if it does not contain conjunctions. If conjunctions are present, the requirement should be broken into two requirements. A point of clarification is in order. A conjunction that is required to define the object of the requirement is valid; however, a conjunction that provides for two different capabilities to be built it not valid in a requirement.

The main point here is that requirements decomposition and derivation is the lifeblood of system development. If the requirement decomposition and derivation is not done correctly, there is little hope of the building a system that meets the needs of the customer. Developing a viable and usable system that meets the customers' requirements doesn't happen by luck EVER.

Question to think about:
Will the requirements for a tier of a system make everyone satisfied?

5.2 Designing for Maintenance

Designing for Maintenance (DFM) is very often overlooked in the overall architecture/design/development activities, and is often skipped because of high priorities like schedule-driven objectives. Regardless of the development style (e.g., agile, spiral, waterfall, etc.), regardless of the mission/business models used to drive the overall architecture and design, building in the notion of DFM from the beginning of the analysis and design process will influence the entire lifecycle process and deliver a system that is maintainable; reducing the overall LCC for the system.

When the system does not have maintainability requirements (e.g., Mean Time to Repair < 8 h) or system engineering delays the inclusion of them into the design, the maintainability concepts, requirements, and changes to System Architecture to accommodate maintenance typically occurs late in the development lifecycle, producing serious system risks, and affecting cost, schedule and technical decisions to make up for the lack of DFM. DFM influences the architecture, design, and implementation of the system for both the hardware and software capabilities by accounting for the maintenance activities on the system. Taking DFM into account early and often in the design will build maintenance into the overall system implementation, installation, and operations; typically reducing the overall maintenance activities, providing maintenance activities which are easier to execute, and reducing the overall logistics (personnel) support time required for system maintenance.

Question to think about:
What happens to a system development if the start-up and shut-down requirements and design are addressed at system integration time?
When should start-up and shut-down be inserted into the development?

What follows is a set of overriding guidelines when building in DFM from the initial analysis and architectural designs. These overall guidelines were created to help the Systems Architect/Analyst begin the process of Designing for Maintenance. For the MDSE, embracing DFM hopefully will result in the following not being a side effect (see Fig. 5.4).

5.2.1 Enhancing System Maintainability

Enhancing system maintainability provides for specifying in the designs and materials such that the system does not encounter unnecessarily and prolonged maintenance activities. This includes the selection of materials used in the design, whether it is considered hardware or software/firmware (whether utilizing COTS or custom). During the Systems Architecture/Analysis, the overall design should take into account the amount of time that would be required to service the system, if these hardware and software components were utilized.

Fig. 5.4 Consequences of not designing for maintenance

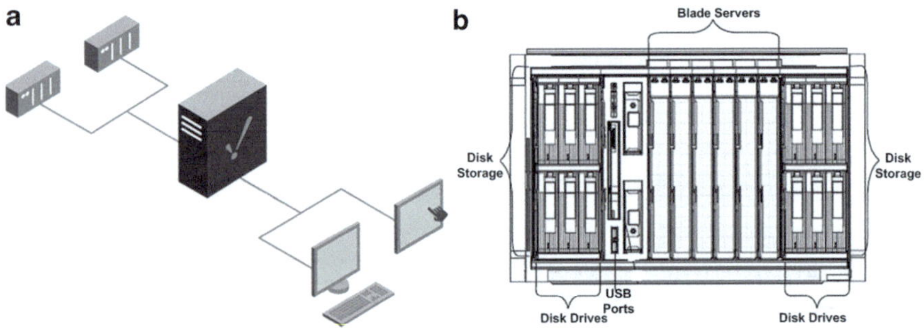

Fig. 5.5 (**a**) Stand alone server configuration, (**b**) blade server configuration

If for example, you have selected a stand-alone computer system that requires manually connecting all of the network and peripheral devices, rather than say, a modular blade server rack with slide in blade servers, you impact the availability of the system anytime maintenance is required (see Fig. 5.5a, b). In the case of software, if you select a COTS software product with a licensing agreement that requires a vendor to be on-site to handle software issues and personnel that operate the system are not allowed to "fix" problems with the COTS software (this is a normal arrangement for COTS SW), either the system would be out of service for a considerable time waiting for a service representative from the COTS SW company, or you would have a considerable cost to have a service representative from the COTS SW company on-site. Such selections should be taken into account in the Operations and Maintenance, including the ease of performing corrective/preventative actions that can be performed by on-site personnel.

5.2.2 *Standardization of Components*

When applicable, it is best to utilize standard components; with the advancement of hardware technologies, and standardization across the manufacturing industry, the use of standard components could simplify the design, but could also tie you into a particular technology; possibly violating certain quality attributes that must be met. If the decision is made to custom build either hardware or software, it increases the development cost due to:

- The design and implementation (which may involve customized parts) of the custom solution.
- Maintenance cycle costs, since personnel must be trained for maintenance of each custom hardware and/or software component in the system
- If there are custom parts, you must ensure you specify sufficient customized parts be inventory for the life of the system.

In the last few decades the use of COTS H/W and S/W has become more prevalent to shorten development cycles and reduce up-front maintenance costs. However, COTS also has certain issues:

- Requires lengthy maintenance agreements with the COTS vendors.
- Obsolescence of the COTS products.
- COTS companies may go out of business.
- COTS companies can get acquired and their product lines can be discontinued.
- COTS solutions can be less flexible as a system grows and evolves while the operating environment changes over time.

5.2.3 *Standardization of Interfaces*

The use of standard interfaces and interface configurations provides similar savings as using standardized components. USB is a standard serial communication interface, which is typically associated with a standard physical component hardware connector. The current version of USB, USB 3.1 specification has a maximum transfer rate of 10 Gbps. USB has standardized serial interfaces. Another standard is TCP/IP, a standard Internet protocol used to define how data is packaged for transfer across the Internet. Again, like standard components, the use of standard interfaces can reduce development and maintenance costs in both labor and materials, and can be utilized to enhance system interoperability across the architecture enterprise. However, it depends on the performance requirements, as there are instances where non-standard interfaces may be required.

5.2.4 *Standardization of Maintenance Manuals*

Ensure that systems are maintainable through easy to understand maintenance manuals and system documentation. The maintenance information should include the required skill levels for the maintenance organization that is responsible for the system. To design a system that requires highly educated personnel to maintain the system, radically increases the overall lifecycle cost of the system. If the system is designed such that a Ph.D. is required to understand the maintenance procedures, then the notion of Design for Maintenance has not been a driving force in the overall design of the system (unless, of course, the system under discussion is the Large Hadron Collider (LHC) used by CERN).

5.2.5 *Ease of Accessibility*

Ease of accessibility provides for sufficient space around maintenance points to ensure access to the equipment should provide sufficient space for safe access. This can be based on designing the enclosures for ease of access, or access panels. When maintenance requires the system to be disassembled to perform routine maintenance, the costs are higher to disassemble and the time of outage is much higher (e.g. replacing spark plugs on an engine where the mechanic can easily access the older plugs take less time and labor costs to maintain, than the engine that requires the air intake and emission systems removal prior to gaining access to the spark plugs).

5.2.6 *Ease of Maintenance Activities*

Design the system in such a way that it can only be maintained the right way—as with access points above, care should be taken to design the system so that disassembly and reassembly of the system results in proper performance. Examples:

1. If the electrical wires [positive, neutral, ground] are important to avoid "*smoking the equipment*", then the plugs and/or adapters should be designed to ensure that the plug cannot be "reversed" when installed.
2. If data loss is not allowed in a system, then redundancy in processing is required such that you can perform "*incremental*" upgrades to the processing services. This then requires the system engineers to address transaction management solutions for data flows in the event the on-line service goes down while the back-up service is in maintenance, to support fault recovery.

5.2.7 Handling of Component Materials

Components that are regularly replaced need to be easy to handle to reduce maintenance labor costs, and safety of both the mechanical aspects of the system and personnel handling the materials. As in the example above on spark plugs, the engine where the mechanic can gain access to all plugs without mechanical change, allows the plugs to be replace with little or no impact on the mechanics of the engine. But in the case where the air intake and emission systems have to be removed, the potential for damage to the engine, and risk to the mechanic goes up as more wear and tear is occurring on the engine during spark plug maintenance cycle.

5.2.8 Designed for Safety

Guarantee safety by the design itself is critical in the design for maintenance of a system. Safety, of both the system and the operators/maintainers of the system, is paramount to system operations. This topical area is subject to volumes of books, regulations, and safety standards and alike across various industries. A good example are the fans used in homes, computers and alike typically have been designed with cages around the moving fan blades, to prevent injury to personnel when the fan is on. While this does not prevent "stupid from happening", the cage design provides built in safety features for the moving blades. In some fans, if the cage is opened, the power to the fan turns off, further extending the safety features of the fan. In some electrical power supplies, the electrical circuits are designed to level load the current entering the power supplies to address intermittent connections and currents on the power lines, so that the swings in power do not burn out the electronics receiving power from the power supply.

5.2.9 Designed for Modularity

Designing modular systems enhances the ability to perform maintenance. Modular designs allows for the software, hardware or both to be replace/upgrade and like capabilities without large lifecycle costs. When a system design has not properly isolated capabilities for maintenance, the entire system is at risk of being down during maintenance, or worse yet, requires a full system retest after a component has received a maintenance update. Modular designs, when properly implemented, allows for maintenance to be performed synchronously to system operations, or reduces the regression testing required on the system before it is placed back into service.

5.2.10 Designed for Failure Modes

Designing the weakest link is important in system design relative to maintenance. Systems engineers should assess where they want the system to fail, and design the component, and access to the components, such that a system failure can be easily addressed by maintenance.

1. In a snow blower, the engine and transmission are critical and costly to maintain and replace; therefore, the designs typically include a shear pin, which will break under heavy load, to protect the engine and transmission from a jam in the show throwing operation.
2. In a communication system, which spans space and ground transmissions, the systems engineer needs to assess where in the communication system the weakest link will occur, and design the system and transition of data to ensure critical information is not lost. A good example for data processing systems is the use of guaranteed delivery (e.g. TCP/IP) vice unguaranteed delivery (e.g. UDP). If the system cannot afford lost data, then it needs to be designed with a guaranteed protocol or retransmission sequences to ensure guaranteed delivery of data.

5.2.11 Designed for Enhanced Reliability

Avoid unnecessary components is exceptionally important to system maintenance. All hardware and software components in the system must provide critical system functions. Unneeded hardware and software components used in a system increases the life cycle costs of the system as well as maintenance costs of the system. Over time, various vendors in the data processing industries have opened their products up to users, which includes a category in the industry know as FOSS (Free and Open Source Software). While it may be free to use to develop with, it is not free from a maintenance standpoint. When software developers bring in hundreds of different COTS/FOSS software products to implement the design, it introduces integration, security and maintenance costs on the system. If the end users don't care about life cycle costs, it may be viable, but when the LCC of the system is important, software should be maintained to the lowest possible number of COTS/FOSS products, and the developers should be required to develop their software to the limited set of COTS/FOSS selected. This increases the reliability of maintenance, as you eliminate the effects of changes caused by the daily update of products in the software industry.

5.2.12 Designed for Enhanced Monitoring

Overdesign system components should be avoided to reduce overall development costs; however, inclusion of key capabilities such as in-bound range checking, fault logging, debug logging, and alike for software increases the likelihood that maintainers can isolate and correct defects in the software. By incorporating monitoring capabilities in the hardware and software designs will increase the fault tolerant behavior of the system, which leads to more reliable and maintainable systems. A cost trade against the objectives of the development must be performed in relation to lifecycle costs including maintenance. The longer the system will be in service, the more reliable the hardware and software needs to be if the objective is a low LCC.

5.2.13 Designed for Expandability

Design for under-stressed use is also a consideration. Low use systems that fail, typically fail when the operators/users of the system cannot afford the failure. If the system is a low event processing capability, whether in hardware or software, the systems engineers need to consider the effects on hardware components that may be idle for long periods of time. In software applications, more robust data handling needs to be considered so that when the software is exercised it is highly reliable.

5.2.14 Designed for Testability

Use components and materials with verified reliability improves the ability to perform Reliability and Maintainability Assessments, such as Mean Time Between Failures, Mean Time to Restore and alike. When hardware and/or software do not have the "vendor supplies reliability metrics," the cost of MDSE for DFM will almost always increase. This is due to the additional analysis and inspection functions that must be undertaken. Once fielded, the expectation for most operational systems is that they fail gracefully and seldom. If the component life expectancy is low, then the systems will typically fail rapidly. A good example is roofing on houses. The single composite shingle roof on a house will fail must faster than the new composite concrete or steel shingle roof. This means that the homeowner has to change the roof more often, which then leads to more damage to the roof supporting material that are constantly receiving new nail holes as the shingles are replace.

5.2.15 Designed for Redundancy

Providing redundancy in system services, as well as hardware, increases the maintainability of the system, as well as the availability and dependability when done properly. Dual paths in the network, processing server clusters, mirrored raid disk drives are all examples of hardware redundancies. When hardware redundancy is employed, automated failover capabilities must be designed into the system to address outages on the primary or secondary devices for maintenance, as well as device failures. In the software arena, service managers provide capabilities to load level process requests, spawn new services to meet increasing demands for peak processing times as well as simple load balancing. These types of software services should also include design for transaction management, to enable fault recovery during startup, shutdown and process crashes.

5.2.16 Designed for Flexibility

Use parallel subsystems and components when you want to improve RMA or flexibility relative to redundancy. Routers in the network are typically paired for redundancy when highly reliable network communication is needed. In some ground antenna, the antennas have dual strings of control equipment for redundancy, and secondary connectivity needs. The use of in-line redundancy or parallel redundancy should be designed into the system so that all considerations for nominal and contingency operations are applied to meet the system objectives.

5.2.17 Designed for Fault Recovery

Distribute workload equally over parallel subsystems or components should be employed any time redundancy is applied to the design. This allows workloads to be shared and bottlenecks minimized during processing. If one service is constantly at 100 % processing and the other picks up the pieces, through put on the system is likely less reliable than when processes are steady state (namely due to the number of transactions that may required fault recovery if the process at 100 % fails).

5.2.18 Designed for Robustness

Design robust interfaces between components/subsystems for reliable processing of information. There are multiple methods for interfaces and increasing the reliability between components. Both hardware and software interfaces should be made to be highly reliable. In older systems, to perform hardware maintenance on devices required the hardware to be powered down, but in new modern hardware components, the hardware is hot swappable. The same is true for software, but in software services, key considerations come into play and should not be over looked, such as, transaction management, in-bound and out-bound parameter checking, fault isolation and recovery and alike.

An important capability that is often overlooking in development, but provides extensive value in development, integration, test and maintenance is unit testing and specifically unit testing that validates the interfaces between components and services. When done properly, each nightly build of software can include auto-regression tested prior to compilation.

5.2.19 *Designed for Environmental Issues*

Design the system to withstand environmental influences typically comes into play for outdoor systems, high altitude systems and/or underwater systems. Mother-nature loves to consume natural and man-made creations. Therefore, all systems should address environmental factors, which are numerous and subject of volumes of design standards:

1. The typically processing environment operates at room temperature (e.g. desk top, laptop…), but in some cases racked equipment has special needs, such as under floor cooling, because of the amount of equipment in an enclosed space. This then requires the racks to have ingest and exhaust venting, which impacts the facility, where raised floors with cooling systems and alike come into play.
2. System engineers should evaluate the environment that the system is delivered into, and ensure that all factors are accounted for in the requirements, design, development and inspection processes. When this does not occur, then rework is much higher as you enter into the maintenance phase of the system's lifecycle. In some cases, the maintenance costs may dwarf the development costs of the system.

5.2.20 *Designed for COTS Management*

Component Management—Minimize the number of, and different kind of, components in the system should be a must as discussed earlier in this section. Somewhere along the path to modernized systems solutions, the COTS market place became very dynamic and robust, leading to large inventories of hardware and software solutions to be available to developers. In many cases, new developments took on a COTS flavor, where the management/procurement organizations pushed new developments into using COTS market place solutions rather than custom developed solutions. When developers use large volumes of COTS/FOSS solutions in their system, they will typically have to spend large around of costs in "glue code" if they want to be flexible enough in the design to replace older COTS/FOSS with new versions. In most cases, the market place of today outpaces the ability of the developers to field new system solutions. This means that newer systems, which are relying more upon COTS/FOSS than custom code, have more system development volatility and maintenance costs. Therefore, systems engineers should work diligently to perform robust trade analyses on each hardware or software system component, to ensure development and maintenance objectives are met. The recommendation for all systems engineers is to strike a careful balance for the number of COTS/FOSS solutions in the design of any system.

5.2.21 Designed for Parts Management

Systems engineers must take spare parts management into considerations when designing the system, to avoid the need to keep expensive spare parts in stock. Engineering the system with spare parts pipelined factored in reduces the need for large expensive inventories, as well as risk to the inventories when mishaps occur (e.g. water leak in supply room fire system). Just in time management for spares should be considered when the spare parts are readily available in the market place. This will then lead to the need to plan for and forecast the use of spare parts and/or maintenance action.

5.2.22 Designed for Equipment Monitoring

Systems engineers should include monitoring equipment in the system design. When you do not build in monitoring functions, then there is no insight to up and coming failure (e.g. automobiles have thermostats, which monitor the water temperature flowing in the engine). When the engine water pump fails, or there is a leak in the cooling system, the thermostat monitoring function warns the operator of impending engine failure, so that they can take action to prevent engine failure. It goes without saying, in order to enhance the ability to operate and maintain a system, engineers must build in monitoring equipment and processes in the system design.

5.2.23 Designed for Prognostic Health Management

Forecasting Maintenance—The design of the system should provide the capability to forecast maintenance actions and/or material consumption to include spare parts. Historically, forecasting system maintenance has significant improved longevity and is an engineering field all unto itself. In the golden days of systems engineer, the *Industrial Engineers (IE)* would evaluate production methods, failure mechanisms, and develop facilities layouts, product workflows, as well as planned maintenance actions into the systems design. In the modern age, there are a host of reasons needed for forecasting maintenance, and these factors should be accounted for in the design:

1. The first step in maintenance forecasting is to factor in the RMA analysis noted earlier in this chapter, and provide a maintenance schedule for both hardware and software generated maintenance need (e.g. Hardware may require down times for lubrication, where software may require disk management scripts run to keep the disk operating a maximum throughput capabilities (e.g. defragmentation)).
2. The second step is to monitor system operations, such as network traffic flows, sub-system/processor loading, disk I/O and similar functions. When system monitoring has been accounted for in the design, the monitored metrics should be made

available to the operators/maintainers as well as the ability to perform trending on monitored metrics. As an example, wear on bearing surfaces leads to increase in heat in the unit; therefore, in addition to monitoring for heat on a device, the systems engineer should include trending for the metric to forecast when maintenance will be needed or spare parts needs to be installed in the system.
3. The older IE practices are slowly fading away, because of the volatility in the new COTS market places, where new hardware and software replace system components faster than developers can field new systems. Because of the volatility, new innovative solutions are warranted and may include learning management solutions to be applied to system monitoring functions.
4. The bottom line is that system engineering should factor in the ability to monitor the systems behaviors, trend the behaviors in order to project when maintenance actions or spare parts replenishment schedules are to occur, or design in a learning management solution to provide automated assessments for the operators of the system.

5.2.24 Designed for Data Management

To provide for effective long-term maintenance on a system, the system engineers should provide a design that allows access to system data saved across time. This includes both system data as well as system monitoring functions and metrics. System data is governed by the market place, rules and regulations and in some cases law. Systems engineers should conduct trade analysis on all data types, to address both contractual and legal requirements, as well as value to system maintenance.

Storage of highly reliable hardware systems (e.g. space applications) data, provides a substantial benefit to the operator and maintainers of systems when the system is in service for a long period of time. As an example, storage of all test data on a satellite may provide benefit to the operators of the system, if the satellite develops an anomaly. The historical data when compared to the current anomaly data may highlight an anomaly vice a "feature" of the vehicle (Feature is this case: not working as expected, but not broken) [82].

5.2.25 Designed for Tools Standards

The design of the system should factor in the use standard tools (both H/W and S/W) where applicable verses development of custom solutions. This allows maintenance personnel to address one or more systems in without needing specialized training. A good example is the use of HP OpenView for monitoring COTS equipment, or RS-232 traffic monitors for communications traffic. This enhances maintenance across the new system, as well as integration into existing computer centers. When there are non-standard monitoring solutions or maintenance tools, the cost and risk to maintenance increases.

5.2.26 Designed for Staffing Considerations

Maintenance Staffing—The design of the system should address the type (skill sets), quality (training levels) and number or personnel required to perform maintenance on the system. System engineers should ensure that the design of the system and its components require as few maintenance personnel as possible when maintenance actions are performed for any given maintenance task. This affects the cost of the maintenance program in both size (number of maintainers) and training (skill sets) required to operate/maintain the system.

As our systems become more COTS based and standards based, the desire to reduce the maintenance work forces has increased across the enterprises. Personnel with a variety of skill sets should be able to execute maintenance.

5.2.27 Designed for Maintenance Documentation

Systems Engineers should ensure that the system is maintainable by a wide variety of personnel. As such, system documentation as well as maintenance manuals (COTS and Custom developed manuals) should be provided with all delivered systems. These manuals should be concise as well as easily understood by the target operator/maintainer work force. The manuals cover the entire scope of the system, with detailed work instructions as appropriate for planned system maintenance functions. These materials should be validated prior to deployment of the system, or training of the operator/maintainers of the system.

Question to think about:
What is the relationship between DFM and the LCC of the system
How does DFM really relate to the overall system design?
When you do not perform DFM, what system criteria has been left out of the development?

5.3 Designing for Test (Test-Driven Development)

Test-driven development is related to the test-first programming concepts of extreme programming, begun in the mid-1980s, but more recently has created more general interest in its own right [83]. Programmers also apply the concept to improving and debugging legacy code developed with older techniques [84].

Designing for Test in a software development process that relies on the repetition of a very short development cycle have the following steps: first the developer writes an (initially failing) automated test case that defines a desired improvement or new function, then produces the minimum amount of code to pass that test, and finally the new code is refactored into acceptable standards. Kent Beck, who is credited with having developed or 'rediscovered' the technique, stated in 2003 that TDD encourages simple designs and inspires confidence [85].

Questions to think about:
What is the relationship between TDD and System Integration?
Does TDD increase or decrease LCC?

5.4 Designing for Integration (DfI)

Systems Engineers must factor in system integration as part of the development cycle of the system. Key integration requirements:

- Specify specific details of the data to be provided throughout integration, by the both the client and receiving systems.
- For the receiving system, specify the actions that must be taken upon receipt of that data (Often this is handled through a performance document or a written control sequence). This should include boundary condition to be tested during integration.
- Identify the physical locations where this intra-system communication is going to occur and the communication methods required by the integration drivers available. In some cases, development of the test drivers for external systems must be implemented.
- Specify the cabling media required by the communication method. This can include both physical as well as wireless methods.
- Determine a testing method to ensure successful implementation of the designed integration. Again, the test method should include automation when the development periods have multiple iterations [86].

Things to consider:

- An understanding of the availability of the integration drivers will lead the project team to a better understanding of how to structure the procurement and implementation.
- Competitive considerations between manufacturers may limit the integrator to using specific products because of the lack of available integration drivers.
- Who provides the labor to implement the interconnection of the systems using the drivers?
    - This requires an understanding of the expertise available in the local marketplace where the project is being built.

5.4.1 Standards-Based Integration

- The standards-based integrated solution is a high-level integration between disparate systems using building industry-recognized standards-based communication protocols. This approach is very similar to the driver-based approach. The difference is the change from a proprietary driver to a standards-based driver as the communication method between disparate systems. Considerations which should be included in the integration approach are:
- Specify specific details of the data to be provided through the integration, by the sending system
- For the receiving system, specify the actions that must be taken upon receipt of that data (Often this is handled through a performance document or a written control sequence.)
- Identify the physical locations where this intra-system communication is going to occur and the communication methods required by the standard protocol
- Specify the cabling media required by the communication method
- Determine a testing method to ensure successful implementation of the designed integration
- Specify the specific standard communication protocol to be used.
- Define the logging functions that should exist in the system in support of integration, which would also solves DFM issues.

Question to think about:

How does Design for Integration affect the cost and schedule of the program? What is the relationship between DfI and DFM?

5.5 Designing for Operations

All systems have to be designed for operational use. The systems may be autonomous, semi-autonomous, manned, unmanned, self-learning, etc. As such, systems engineers must address all of the factors associated with the planned operational use of the system. Without understanding the operational considerations, system engineering cannot hope to understand:

- How end users' jobs will change with the implementation of the new product or product increment. An evaluation of the extent of the change the new system introduces must be factored into the system such that the operators of the system can adapt to the change. When this is not performed properly, there is an increase in risk for the lack of buy-in, unhappiness, and even possible sabotage by the operators
- Which of the issues associated with the "as-is" system must be addressed. If we implement new processes and systems without curing existing ills, the project will lose credibility, or worse, be rejected.

- No new product or solution is devoid of the current state. Even when new systems, products or services are created, they must fit into the current environment and infrastructure. That means systems engineers have to address requirements that account for the environment the new solution will be deployed into, such that it contains or bridges existing functionality, and without which the new product cannot perform adequately.
- In addition, most projects are not entirely new products, but replacements and enhancements of existing systems or processes. These efforts need plenty of analysis of the "as is" state and its corresponding requirements to make sure the changes developed will fit into the rest of the business operation. This is typically referred to as backwards compatibility.
- Even for brand new products, there are many business rules and decisions that apply across projects that should be taken into account for the project at hand. For example, a new Training Management System built for a company's web site was contracted to a development company. The "specification" the development company used represented the stakeholder's business model requirements. The specifications were clearly not a design—that was done by the developers—but represented the company's needs and requirements for the new system relative to the target audience that the training was developed to support.

5.5.1 Operational Suitability: Ease of Use

Most developments typically have a target audience or users that operate the system. Because of this, most developments include human factors in the system design. The human factors may be represented in requirements, but typically are developed and include in a Human-Computer-Interface (or Human System Interface) standard. The standards typically address factors such as the use of colors, data formats, display operations, etc. In some cases, governmental laws come into play to address personnel with handicaps, such as color blindness, hearing impairment, physical impairments and alike. In all cases, the goal of HCI engineering is to ensure ease of use by the personnel that operate or use the system.

A case in point, a new system was deployed to replace an older system. The time for an operator to enter data in the old system was on the order of 5 s for an alert. The new system came along, using windows, pull-downs and sub-windows, such that the time to enter data for an alert took 3.5 min. The new system capabilities went unused by the operators, and the older system was required to remain on line indefinitely. One may ask why this happened. The answer was quite simple. The number of events occurring in an 8-h shift resulted in 10.2 actual hours of operator time to complete the shift workload. Specifically, the new system was unsuitable for system operations, because the operator became the processing function for use of displays and not the user of the system.

It is very normal for systems engineering to have trained human factors personnel develop the HCI standards, as well as program guidelines for HCI. Typically the HCI standards are a standalone document, or incorporated into the Programming Practices, Standards and Conventions (PPS&C) used by the software developers.

And lastly, while the prior paragraphs address software, HCI incorporate physical workspace required into the system specifications in order for humans to operate and maintain the system. HCI typically includes access to and workspace sizing, terminal layout, phone locations, etc. Maintenance considerations include the use of displays in racks, and how the functional is this relative to the maintenance staff. When done properly HCI engineering improves operational suitability for all systems, to include unmanned systems, because somewhere along the way, the unmanned system has data outputs for diagnostics and like.

Question to think about:
What happens to the delivered system if operational suitability is poor?

5.6 Case Study: Designing for Success That Ends in Potential Failure

The use of blade servers has become commonplace in today's IT-intensive systems. Blade servers provide high processing capabilities with a reduced floor footprint. The use of a blade server chassis reduces and simplifies cabling, storage, and maintenance. COTS software exists to allow load balancing and failover capabilities for blade servers, making them an excellent candidate for Grid Computing applications. Many commercial and defense-related companies utilize IBM Blade Server technology. Recently we saw the following headline:

5.6.1 Sale of IBM Blade Server Division to Lenovo (A Chinese Company)

A Chinese company now owns the IBM Blade Server line. For Department of Defense users, this is disturbing, since American companies must own Hardware and Software utilized on DoD contracts, unless allowed via waiver. Coupling the above headline, with this one:

5.6.2 Superfish Spyware Software Discovered Installed on Lenovo PCs

Such software allows every move online to be tracked, and makes the computers vulnerable to hackers. The combination of these two makes the utilization of, now Lenovo, Blade Servers suspect. Systems that are implemented with older IBM Blade Servers now have serious decisions to make when it comes time to a technology refresh. In some cases, they will not be allowed to refresh the system technology with Lenovo Blade Servers that are compatible with their older servers, and to change out all of the servers to a different manufacturer (say HP or Oracle) requires a complete change out of chassis, and require a complete reassessment of size, power, weight, and other factors surrounding the new grid computing installation. This could radically change the Lifecycle Cost profile of the system. If the choice is made to change out the current Blade Servers with Lenovo Blade Servers there must be consideration made as to possibilities posed by the risk of spyware, even though the discovery was initially made on Lenovo PCs. Lenovo has since fixed the mistake of their PCs being sent out loaded with Superfish, and there is no indication that Blade Servers would suffer from the same issue.

The overall point is not to say that there is an inherent issue with Lenovo Servers, or Lenovo computers in general (or any other computer brand). The point is that the use of COTS, while alluring, has its own set of potential issues that can cause major problems in the overall design, implementation, and/or lifecycle costs of a system.

5.7 Discussion

Not taking into account all of the disciplines required for a System of Systems design results a delivered system that is deficient in one or more areas. Operationally viable systems that meet all of their requirements, objectives, constraints, etc., does not happen by accident and cannot be bolted on in the middle or end of the development life cycle. This chapter addressed Design for Maintenance for a very specific reason. For a system to operate, it must be integrated, test, be fielded and used. When any aspect of DFM is missed, such as startup/shutdown, the system will not operate, or incur a serious risk impact on the development, typically in large cost and schedule overruns. Coding capabilities, or building hardware, where DFM is not factored in is easy to do. But to field a viable operational solution requires all DFM actions to be considered early and often in the development. It is the MDSE team's responsibility to ensure all aspects of the System of System design is folded in at the beginning and throughout design, development, transition, deliver, operations, and maintenance. And, pay attention to risks as they rarely go away by themselves.

Chapter 6
Systems of Systems Architecture Design

According to the IEEE standards, System Architecture is defined as:

> A generic discipline to handle objects (existing or to be created) called "systems," in a way that supports reasoning about the structural properties of these objects [87].

Depending on the book you read, or the context in which a Systems Architecture is utilized, you could also define it as:

- A model to describe or analyze system behavior.
- A method of create the architecture of a system.
- A discipline utilized to create systems architectures, including a body of knowledge for architecting a system in such a way to meet specific mission/business needs.
- The concepts, frameworks, principles and practices, tools, methodologies, and heuristics required to create the architecture, based on a set of given requirements.

This points to the system architecture as a global model of given systems, consisting of:

- A structural model of the system
- A logical model of the system
- A list of properties the system must obtain
- The system behaviors (both static and dynamic)
- Views that illustrate various characteristics of the Systems Architecture (e.g., Service Views, Data Views, Operational Views, etc.)

For the present, Systems Architectures are constructs, designed by people, which involve various system components required to build a functional system. This may include, but is not limited to, hardware, software, networks, operators (human or robotic), processes and procedures, which are orchestrated together to create system elements that are created to perform specific mission/business needs, goals and requirements. The complexity of the system depends on the number of elements, the

J.A. Crowder et al., *Multidisciplinary Systems Engineering*,
DOI 10.1007/978-3-319-22398-8_6

functionality of each element, the integration of the elements, the number of internal and external interfaces, and the heterogeneity of components within the overall system and within each element. The number of architecture levels also determines complexity, as it demands a recursive integration process. The heterogeneity of functionality may require the Multidisciplinary Systems Engineer to bring in a host of specialized fields in the design of extremely complex SoS architectures. The more complex the architecture, the more difficult it is for the Multidisciplinary Systems Engineer to keep a unified vision of the overall SoS and to manage the design and implementation of the SoS.

Question to think about:
Are the levels of the architecture important to the MDSE?

For the System of Systems (SoS) Multidisciplinary Systems Engineer (MDSE), the focus of the architectural design requires an understanding of how the methods and structures of all SoS capabilities, needs, goals, and quality attributes combine to support the overall SoS Architecture. As discussed in previous chapters, each SoS is unique, with different elements, element functionality, requirements, interfaces, and emergent behavior. The unique characteristics of each SoS affects the way Multidisciplinary System Engineering is applied to the SoS Architecture. One particular example is the software-intense SoS. Here, the SoS often suffers from major integration and operational/behavioral problems due to element integration where there are consistency issues between system and software architecture and their designs for addressing the system quality attributes of each element, and for the higher-level SoS. These issues generally result in needing to re-architect and redesign the elements, as well as dealing with the operational errors resulting from the inconsistencies between element designs. This results in major cost, schedule, and mission/business effectiveness.

Question to think about:
Can you integrate a SoS if an architecture tier is missing in the analysis?

For most systems, the SoS architecture drives a software architecture that must be developed and tested. Figure 6.1 illustrates the Software Engineering development lifecycle shown in Chap. 1, Fig. 1.7 being fed by, and feeding, the SoS architecture process.

6.1 System of Systems Complexity

The concept of System of Systems provides architectural methods and constructs to create very powerful and adaptable systems that can operate in extreme and dynamic environments. At the same time, the increased complexities necessitated

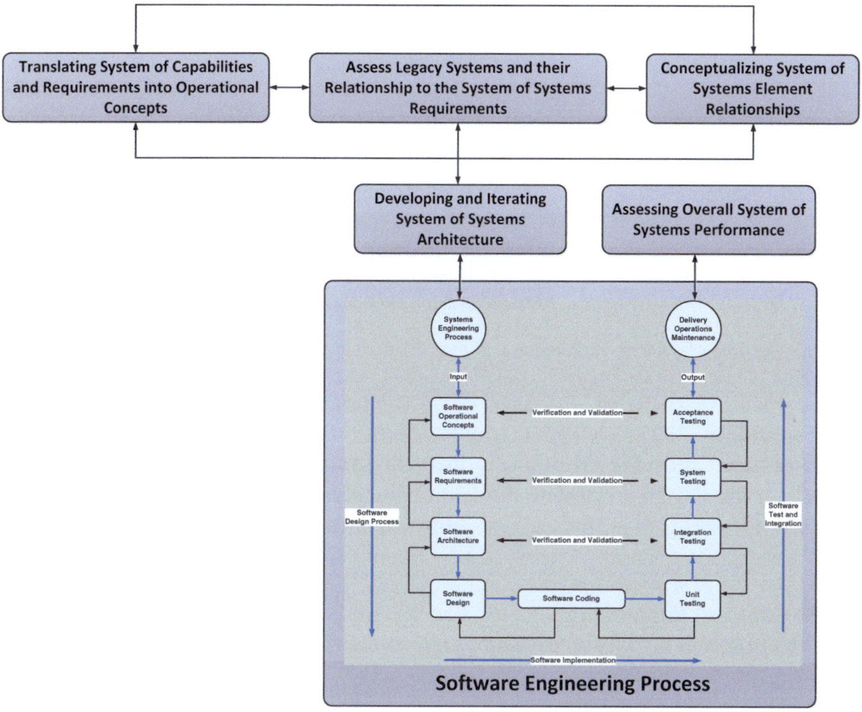

Fig. 6.1 High-level SoS architecture process feeding the software engineering process

by cross-element associations, integration, and increased inter and intra element interfaces creates the potential for issues related to emergent behaviors. As our systems become larger and their capabilities are increased, they add complexity and difficulty in creating the architectures required to fully realize and manage the required System of Systems that meets all of the requirements, performance metrics, goals, quality attributes, etc. This is partially driven by the interactions and effects (dependencies) that each architecture driver has on the overall characteristics of the System of Systems. Figure 6.2 below illustrates these dependencies.

While a well-designed, static architecture may have attributes such as availability, reliability, maintainability, and others that come from a well-structured system, such architectures may lack flexibility, adaptability, and evolvability; attributes that most modern System of Systems requirements have as necessary quality attributes. For the Multidisciplinary Systems Engineer, the design of System of Systems architectures should result in a forward-looking system that can grow and evolve as the needs and environments change over time; adapting to new technologies and/or operational concepts. This should include a new architecture documentation system that allows easily adapted systems and software views required to evolve the System of System designs. We will discuss this much more in later sections, and

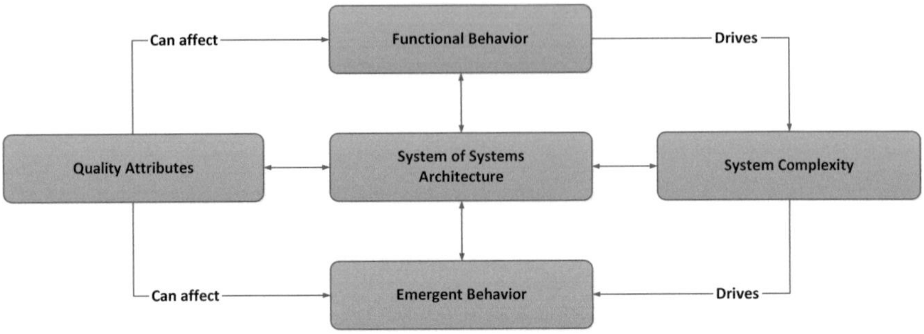

Fig. 6.2 System of systems architectural characteristic dependences

will introduce the Electronic Engineering Notebook and the notion of a Functionbase.[1] For the Multidisciplinary Systems Engineer, understanding all aspects of the design are essential, including analysis and the characteristics of the System of Systems disturbances caused by the infusion of new technologies.

Questions to think about:
If the program decides to avoid an architectural tier to save on cost:

(a) *Does the decision actually save development cost?*
(b) *Is the total LCC actually reduced?*

6.2 System of Systems Enterprise Architecture

The study of System of Systems Enterprise Architectures is an emerging discipline that is focused on improving the performance of Enterprises, analyzing them in a holistic and integrated view of their strategic direction, business practices, information flows and technology resources. By analyzing and visualizing current and future versions of this integrated view, the Multidisciplinary Systems Engineer can better manage the transition from current to future operational modes and methods [88]. This transition includes the identification of new goals, activities and all types of capital and human resources (including Information Technology) that will be required to improve financial and mission/business performance (strategic vision). The follow discussion explores how to visualize and analyze System of Systems architectures, and how different views (slices) through the architecture allow you to ensure that all requirements are captured within the design. The purpose of this section is to provide an understanding of the principles and practices of architecture analysis and visualization.

[1] Functionbase is an electronic documentation methodology which captures the systems, software, and test architecture and design, as well as the context for the design in one hypermedia reuse package.

6.2.1 System of Systems Modeling and Simulation

As discussed in previous sections, the systems analyst confirms that the designed system meets the customer's requirements. Typical analyses would include power, throughput, interfaces, and hardware sizing. Usually, the more complex parts of the system need to be modeled in order to demonstrate that they will perform properly and that their interfaces are correct [89]. Modeling also helps the Multidisciplinary Systems Engineer understand how the System of Systems will be operated. The questions are: How can the System of Systems architectures be adequately represented? How many items or characteristics need to be in the model?

Some relationships between elements and nodes of the System of Systems are static (always exist), while some are dynamic in nature and exist only under certain circumstances (e.g., only appear during certain failure modes). In general, there is no unified representation of a System of Systems that captures both the static and dynamic behavior, as well as capturing both architecture objects and behavior. The following topics provide some of the methodologies utilized to capture and model the characteristics and behaviors of a System of Systems.

6.2.1.1 History of Modeling and Simulation

The science of Modeling and Simulation developed independently in many fields of study (e.g., economics, engineering, sports, etc.), but really flourished and advanced when Systems and Cybernetic Theory became strong disciplines. With the spreading use of computer systems in the last ½ of the twentieth century, Modeling and Simulation techniques and methods began to unify under a high-level, systematic view. There are many types of simulations that can be created, depending on the System of Systems functionality; here we discuss a few of the major categories:

- **Physical Simulation**: this allows designers and engineers to create multiple physical prototypes and computer models for the analysis and design of new or existing physical parts.
- **Human-in-the-Loop Simulation**: in this type of simulation, human operators are part of the simulation in order to handle the human-systems-interface part of the simulation. Most include computer simulations that model a human operations environment, like a flights simulator or satellite operations center. The simulated operational environment is often called a "synthetic environment." This can be used to determine if the System of Systems architecture, functionality, and operator interactions meet functional, performance, and visual requirements.
- **Failure Mode Effects Analysis (FMEA)**: this refers to a simulation to recreate potential system failures and understand how the System of Systems will handle various internal and external anomalies and errors.

In general, given that all System of Systems have certain characteristics, i.e., properties, behaviors, and objects, what methodologies are available for modeling and/or synthesizing these System of System characteristics? To answer this question, a few definitions are required for understanding:

- **Modeling and Simulation**: Modeling and Simulation is the process of designing a model of a System of Systems and its elements, and conducting time-based simulation experiments in order to understand the behavior of the System of Systems and Element interactions in order to evaluate various strategies for the architecture of the system. Modeling & Simulation (M&S) is an important tool in all phases of the System of Systems life cycle and processes, including requirements analysis, architectural design, design & development, V&V, operations and maintenance. Some of the challenges with Modeling and Simulation of a System of Systems include:
  - Multidimensional Trade Spaces
  - Predicting performance across a multitude of design and technology options
  - Performance characterized by multiple Measures of Effectiveness (MOEs)
  - Interdependencies of technologies, operations, and procedures
- **System of Systems Models**: abstract representations of the System of Systems architecture that are used to assess and/or predict the system behavior.
- **System of Systems Modeling**: the methodologies required for creating and testing System of Systems execution models.
- **System of Systems Simulation**: The time-based representation utilized to execute the System of System models, either statically or dynamically (time-varying).

The overall purpose of Modeling and Simulation is to capture [90]:

- Structure and characteristics
  - Non-linearity
  - Hierarchical representation of decomposition
  - Dynamic Behaviors
  - Ease of Creation
  - Ease of Validation and Verification
- Relationships between Elements
  - Adaptive Behaviors
  - Emergent Behaviors
- Interactions with each element and interface (both internal and external). This often leads to a measure of uncertainty in understanding the System of Systems functional behavior.
  - Includes Intelligent Agent interactions if they exist in the System of Systems.

Modeling Techniques that exist to capture the System of Systems characteristics include:

- Case Studies
- Computer Models
  - Optimization Models
  - Econometric Models
  - Simulation Models
    - Network Analysis
    - Markov Simulation
    - Petri Net Simulation
    - Discrete Event Simulation
    - Dynamical Systems Analysis
    - System Dynamics Simulation
    - Cellular Automata
    - Agent-Based Simulation

Table 6.1 below illustrates the relative usefulness of many of these techniques at modeling and simulating various System of Systems characteristics.

Discussed earlier was the notion that various "view" of the system are required in order to communicate the SoS architecture to the MDSE team, but also to the SoS software, V&V, Integration and Test, Deployment, Operations and Maintenance teams. We now illustrate several of the most useful architecture views, the Use Case, the Activity Diagram, and the Sequence Diagram.

6.3 Use Cases

The Use Case, also called a User Scenario, is a construct for capturing how users (human or other processes) will interact utilize the system (interact) by exercising the hardware and software that implement the system requirements. There are two types of Use Cases:

- **A word description Use Case**: represents a brief User Story that describes (in words) who is using the system and what the User is trying to accomplish (see Table 6.2). The User may be a person, processes, or device.
- **A Use Case Diagram**: is a visual behavior diagram which describes the functional requirements exercised by a particular use of the system. The Use Case diagram illustrates the relationships between actors (elements representing the role of person(s), processes, or devices) and the system. Figure 6.3 illustrates a Debug and Logging Use Case Diagram.

In general, word descriptive Use Cases have many elements, as described below in Table 6.3.

Table 6.1 Modeling/simulation techniques vs. system of systems characteristics

System of system characteristics →	Non-linearity	Element interactions	Intelligent agents	Hierarchical representations	Emergent behavior	Adaptive behavior	Dynamic behavior	Ease of creation	Ease of V&V
Modeling techniques ↓									
Network models	5	2	5	2	4	5	4	1	2
Discrete event simulations	2	4	4	3	4	4	2	4	3
System dynamic models	2	4	74	3	4	4	2	3	3
Agent-based models	1	2	1	1	1	1	2	5	5

Excellent = 1, very poor = 5

Table 6.2 Use case example

10.	This flow begins when a new User requests access to the
10a.	An account request form is completed with User information, requested roles and accesses, and supervisor approvals
10b.	The system administrator verifies that the new User has the appropriate need for the system roles requested
20.	The system administrator conducts related system use training to the new User
20a.	General User: training covers information security rules and regulations, the expected behavior of Users, and the consequences of non-compliance
20b.	Specialized User training: training for individuals who have specific needs for system information (e.g., User Credential)
20c.	Users are required to sign an acknowledgement of the rules and regulation before access is granted to the system
30.	The system administrator and supervisor approves the account request, including requested roles and privileges
40.	The system administrator creates the new account. Only system administrators are authorized to perform account management functions
40a.	The system administrator assigns a unique User ID and creates an initial password. Upon initial logon, the User will be required to change the password following password policy guidelines
40b.	The system administrator will assign a User certificate for authentication and authorization purposes
40c.	Approved User roles are added to the User Accounts based on their authorized duties and roles. Each role will give the User certain permissions and access to perform authorized tasks. Users may be granted more than one role

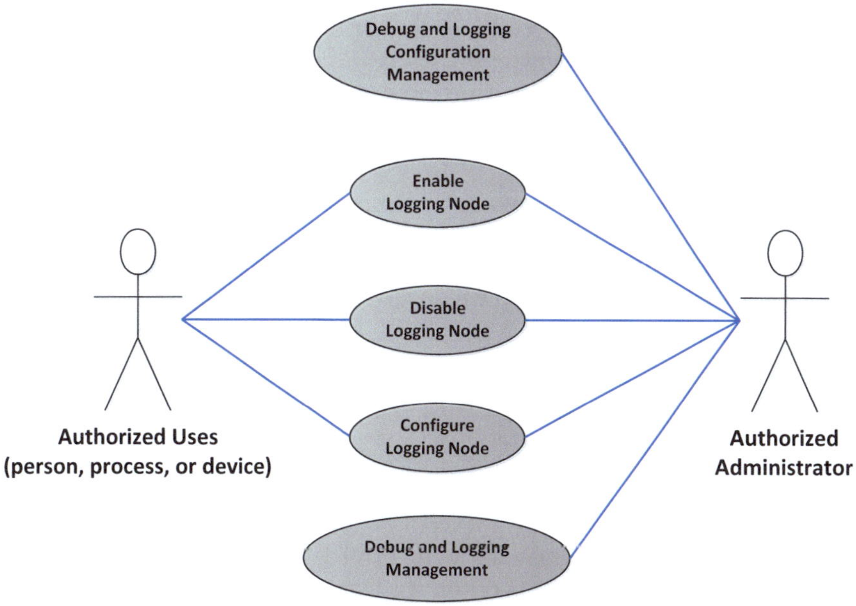

Fig. 6.3 Example use case diagram: debug and logging

Table 6.3 Use case elements

Use case element	Element description
Use case number	ID number used to identify the use case
Use case application	What is the system, element, subsystem, application, etc., for which this use case pertains?
Use case name	Title for the use case. The title should be short and descriptive; should represent the goal the use case is trying to satisfy
Use case description	What system, element, subsystem, etc., functions is this use case executing; done in paragraph form
Use case primary actor	Who is the main actor for this use case (person, process, or device?)
Preconditions	What preconditions must be met before this use case can start?
Trigger	What event(s) trigger this use case
Basic use case flow	The basic flow should describe the events of the use case when everything is perfect (happy path); there are no errors, no exceptions. The exceptions and errors are handled in the "alternate flows."
Alternate flows	Significant alternatives, exceptions, and errors handling

For Use Case diagrams, a simplified description of rules for creating a Use Case Diagram is given below. This, and the following sections, is not intended to be detailed descriptions of the architectural methods and structures, but rather an introduction to these concepts. There are many websites and books available for each of these architecture description methods:

- A use case describes a sequence of actions that provide something of measurable value to an actor and is drawn as a horizontal ellipse.
- An actor is a person, organization, or external system that plays a role in one or more interactions with your system. Actors are drawn as stick figures
- Associations between actors and use cases are indicated in use case diagrams by solid lines. In Fig. 6.3 note the associations between the actors and the actions described in the ellipses.
  - An association exists whenever an actor is involved with an interaction described by a use case.
  - Associations are modeled as lines connecting use cases and actors to one another, with an optional arrowhead on one end of the line.

6.4 Activity Diagrams

Activity Diagrams are utilized for mission/business process modeling, and normally are utilized to capture the logical flow for a single Use Case or mission/business rules. Activity Diagrams are useful to model the internal logic for complex operations, but it is often easier to rewrite the Use Case or mission/business rules to be simple enough to not require an activity diagram to understand it.

Activity Diagrams equate to object-oriented data flow diagrams from older structured development architecture design methods. The components of an activity diagram are described in Table 6.4 below.

Although Activity Diagrams are normally used to depict the activities across a single Use Case, that can also be drawn to address multiple Use Cases and describe a group of activities constituting a process thread through the architecture, or can be used to depict just a small portion of a single Use Case, such as an alternate flow or error handling.

Activity Diagrams are also utilized outside of Use Case depiction; often used by Agile or eXtreme programming teams to draw Activity Diagrams to analyze a mission/business process to be coded during the current Sprint [91]. Activity Diagrams, as can be seen in Fig. 6.4 can become complicated. Figure 6.4 depicts an Infrastructure Activity Diagram which is utilized by the entire System of System Enterprise, and therefore there are actors; every person, process, or device within the System of Systems makes use of Transaction Management.

Table 6.4 Activity diagram elements

Activity diagram element	Description
Activity initial node	Small, filled in circle that is the starting point for the diagram. The initial node is not required, but makes the diagram easier to follow
Activity final node	Filled in circle with a border is the ending point. The activity diagram can have zero to many final nodes
Activity	Rounded rectangles represent activities that occur, based on use case steps. And activity may be a physical activity, or an electronic activity, and are accomplished by a person, process, or device
Activity flow	The arrows on the diagram that indicate process or information flow through the activity
Activity fork	A black bar with one flow entering it and several flows leaving it. This denotes parallel activities
Activity join	A black bar with several flows entering it and one flow leaving it. This denotes the end of parallel activities
Activity condition	Text on one of the flows which defines a condition which must be evaluated and must be true in order to traverse that node
Activity decision	A diamond with one flow entering and several flows leaving the diamond. The flows leaving the node include conditions required to traverse each of the flows
Activity merge	A diamond with several flows entering and one flow leaving. All of the flows must reach this point before the overall activity flow can continue
Sub-activity indicator	A more finely detailed description of an activity. A sub-activity is indicated by a rake in the bottom corner of an activity
Final flow	A circle with an X through the circle indicates that the activity process ends at this point

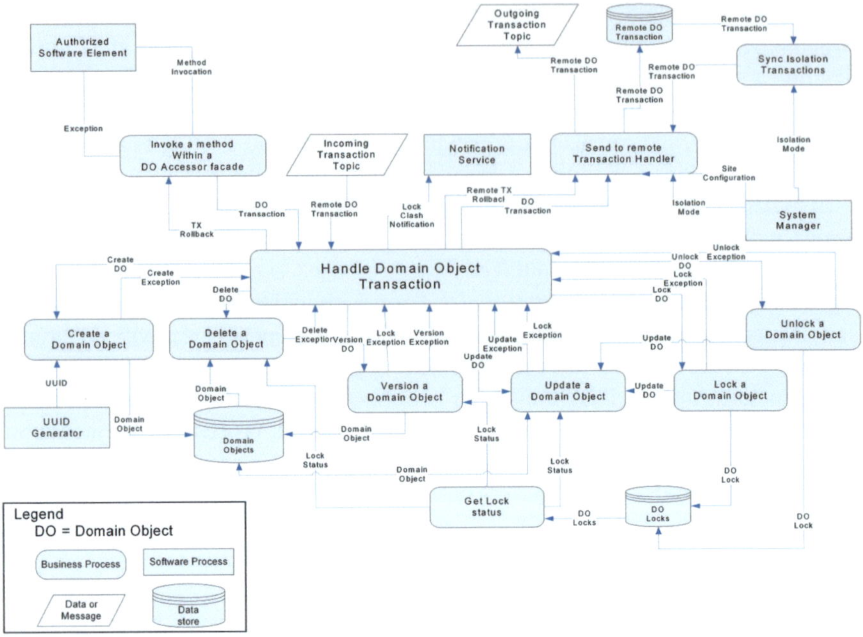

Fig. 6.4 Sample activity diagram: system of systems enterprise transaction management

6.5 Sequence Diagrams

While Use Cases provide a useful look at what the system is supposed to do, or the steps that should be taken in execution of system functions, it does not provide information on how those steps are to be executed, the data/information required to perform the Use Case steps, or the sequence of events that must take place in the execution of the Use Case steps. This information is provided through the use of Sequence Diagrams. Sequence Diagrams models the flow of logic and information within the system in a visual matter, providing information on how objects, actors, and other processes within the system collaborate, based on a time sequence; based on a given Use Case or system use scenario. Basically, the Sequence Diagram enables the Multidisciplinary System Engineer to document and validate the flow of logic through the System, Elements, Subsystems, Services, Components, etc. within the System of Systems. Sequence Diagrams are useful for modeling the dynamic behavior of the System of Systems in order to validate the overall architecture design [92]. Sequence Diagrams are useful modeling tools for:

- System Scenarios or Threads: A system thread is a description of a potential way the system is used.
    - The logic of a System Thread may be part of a Use Case, or provide an overall single usage throughout the system, or potentially an alternate course of action, caused by error conditions.

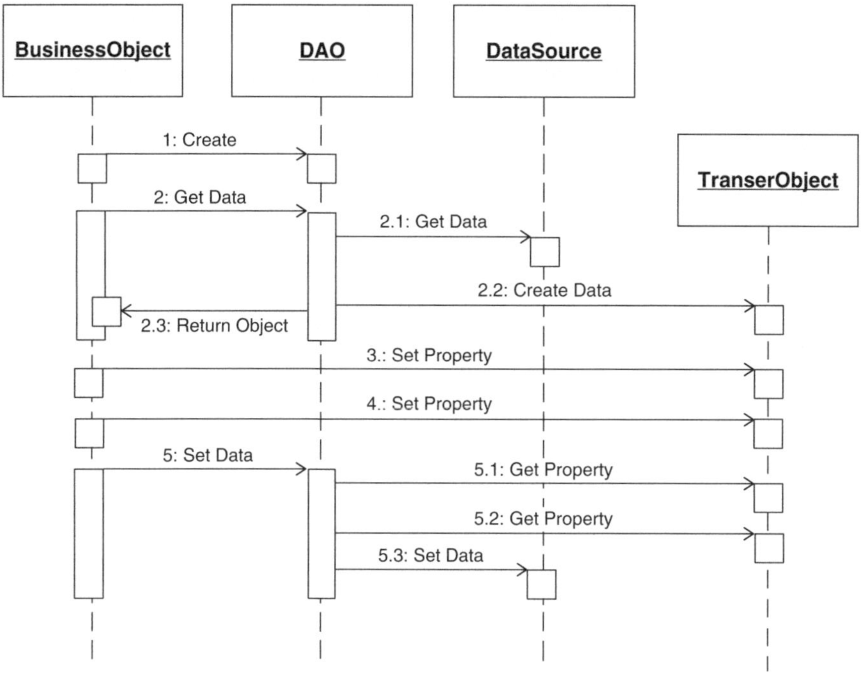

Fig. 6.5 Sample sequence diagram: data access object creation and retrieval

   - The logic of a System Thread my pass through the logic of many Use Cases and trace the logic throughout the system for a single purpose.
- Sequence Diagrams are useful to trace the logic and information flows of a complex operation or function, providing a visual object code through the System.
- Sequence Diagrams are popular for modeling dynamic behavior; allows focus on identifying the behavior of functions within the System of Systems.

Figure 6.5 is an example of a sequence diagram providing data access services within the System of Systems Enterprise Infrastructure.

6.6 Architecture View Discussion

There are many more UML diagrams that can be utilized to capture different aspects of the overall System of Systems architecture. These include, but are not limited to: Communications Diagrams, Class Diagrams, Entity Relation Diagrams, Timing Diagram, Deployment Diagram, and State Diagram. In addition, there are a host of architecture views we have discussed from DoDAF, MODAF, and TOGAF (e.g., Service Views, Operational Views, etc.). Each of these architecture constructs provide a different view into the architecture and design and may be useful. Figure 6.6

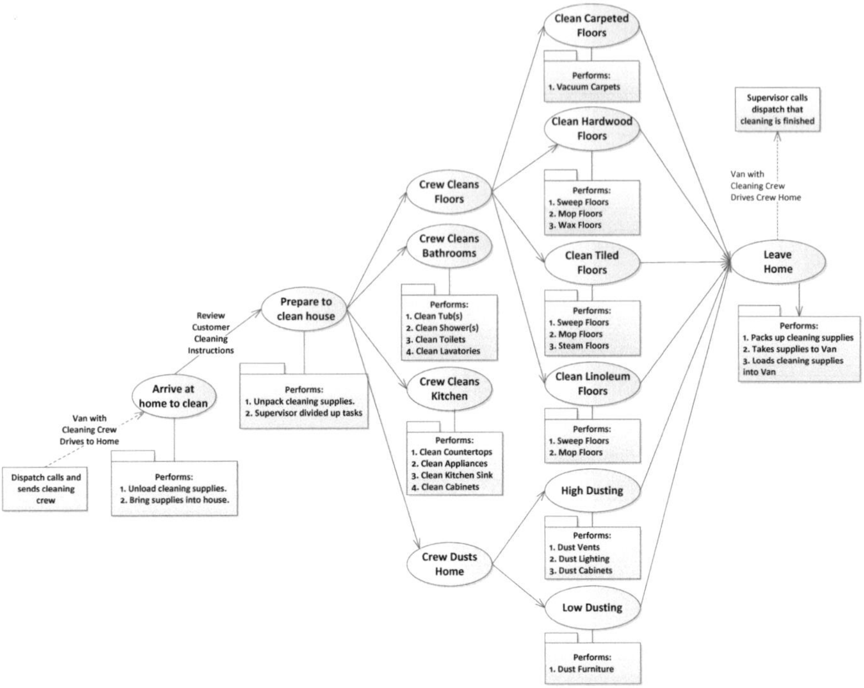

Fig. 6.6 OV-2 for a house cleaning service

illustrates one of the Operational Views, the OV-2 (Operational Node Relationship Description). All three of the architectural frameworks discussed (DoDAF, MODAF, and TOGAF) have an OV-2 as part of their architecture views. The OV-2 defines the nodes which provide a description of the capability requirements with an operational context and the relationships between the nodes. The OV-2 shown in Fig. 6.6 represents the node description of a house cleaning service, which will be used later to discuss other program architectural representations, such as the Work Breakdown Structure (WBS) and Organizational Breakdown Structure (OBS).

One of the issues that must be decided is what level of detail is required for a given system architecture. Very few system architectures require all of these views/diagrams. It is up to the Multidisciplinary Systems Engineer to review the requirements, CONOPS, and other documentation to determine what views may be appropriate for given design. The next section discusses the problems with taking architecture diagrams too far.

6.7 Architecture Pitfalls: An In-Depth Discussion

Any Multidisciplinary Systems Engineer has access to more architectural design techniques that they can possibly use, or afford to apply to any given project. The issue is deciding how much is enough? The tendency is more often than not to apply more views and types of views than the design project warrants; managers ***LOVE*** documentation. The main thing to keep in mind is prominence. While architectural design techniques can be applied to any project or project format (e.g., waterfall, agile, eXtreme, etc.), the key to a successful project is choosing from the alternate designs, eliminating those that are doomed to fail, and finding those architectural constructs and designs that minimize risk for the given project. This is not always possible, as any challenging design poses problems and issues that evolve as the project progresses, and the Multidisciplinary Systems Engineer must tackle high-risk designs, moving forward with designs that may or may not work. Historically, engineering as a whole can be viewed as Henry Petroski [93] does:

> *The concept of failure is central to the design process, and it is by thinking in terms of obviating failure that successful designs are achieved. Although often an implicit and tacit part of the methodology of design, failure considerations and proactive failure analysis are essential for achieving success. And it is precisely when such considerations and analyses are incorrect or incomplete that design errors are introduced and actual failures occur.*

The questions that must be answered by the Multidisciplinary System Engineering Architect are:

- How much is enough architectural documentation?
- What techniques are applicable to this particular design project?
- How does the overall Systems Engineering organization ensure that the architecture keeps up, or evolves as the requirements change and as the system design evolves through discovery of the emergent behavior as the System of Systems design is tested and evaluated?

The answers to these questions are central to Multidisciplinary Systems Engineers in learning how to choose architecture design techniques in order to reduce design and implementation risk across the development efforts, and to balance a well-planned design with a more adaptable, evolutionary design. That balance is difficult and often abandoned for a more traditional approach; one fraught with its own set of difficulties. There are many wrong approaches that have been used over and over throughout the decades, none of which have proven to be very effective:

- **The Ad Hoc Design**: here, engineering reacts real-time to project needs and decides at each increment in the development how much architecture and design to go forward with. This approach is, unfortunately, popular, even though it usually results in a haphazard design that is very subjective in nature; rarely capturing the needs of the customer and/or meeting the mission/business requirements [94].
- **Weighted Design**: in the weighted design, the project is bid on a percentage basis, based on the software to be developed, or the Software Lines of Code (SLOC) that are expected. Everything is based off of this estimate. An example would be 15 % time/money spent of architecture, 50 % on writing software (coding and unit testing), 15 % on integrating the software, and 20 % on system testing. First the cost and schedule estimates for software are computed, and then the other elements of the development (architecture, integration, and system testing) are based off of the software estimates. Invariably, the software estimates are wrong, as they are based on VERY optimistic views of the software; SLOC, efficiency, the number of issues (or bugs) that will be found in the code, etc.

Very few programs bid this way come in on-cost, on-schedule, and on-performance. This methodology helps make planning easier, even though the notion that for every system architecture design you can utilize the same weightings is very short sided; basing everything on the SLOC is equally problematic with today's modern languages, coding techniques, and code generators. SLOC is a bidding methodology is an outdated construct that should be abandoned.

- **No design**: this may seem similar to the Ad Hoc style, but it is different. Software Coding and Unit Test (CUT) is started immediately, based on the experience of the software staff to "know what to code." Here, the design eventually happens, but is an inadvertent result of coding and happens at the computer workstations, rather than any up-front architecture design. This methodology is particularly high-risk, since there is no guarantee the software will somehow evolve into an integrated, coherent system that meets all of its requirements.
- **Building a "complete" architecture documentation set**: Here the system architect employs a very comprehensive set of design documentation that produces a complete written design volume. Unfortunately, these are usually static documents that do not change and evolve as the system does, resulting in a system design that is completely out of touch with the system implementation. This creates a maintenance nightmare, since the system documentation does not reflect the system implementation.
- **Standard Processes/Templates**: employing templates or design patterns is not intrinsically bad, but they are often utilized poorly [95]. A template is not a substitute for analysis and architectural design; more, they are useful to help with the mechanics of documenting the architecture, once an architecture style has been chosen.[2]
- **Using Industry Standards in Non-Standard Ways**: utilizing industry standards in non-standard ways (making up your own terms or views) is counterproductive, since it is very difficult to communicate using standard tools and is almost impossible to ensure consistency across the architecture. Since architecture design tools are created to allow consistency checks using the industry standard terms, views, components, etc. within the tool, using non-standard terms, diagrams, views, etc. Communicating non-standard architectural design artifacts to customers and to the rest of the engineers across the project is very problematic, since each person will bring their own interpretation to each and every non-standard element you introduce into the system.

The bottom-line purpose of the architecture is to communicate it to the rest of the program, to the customer, and ultimately to the operations and maintenance crew. Communicating the architecture adequately is a difficult task, since the term adequate has many, many different interpretations. Here, the Multidisciplinary Systems Engineer must ask some basic questions:

- What architecture design language should be used?
- What visualization medium do the customer and development team utilize?

[2] Sample templates for most plans can be found at: www.ProjectManagementDocs.com.

Just a note: ***POWERPOINT IS NOT AN ARCHITECTURE DOCUMENTATION MEDIUM!!***

Too many projects get wrapped around the axle in arguing about the design language and documentation tools. Too many engineering managers fail to realize that the goal of architecture documentation is to communicate the architecture for buy-in/consensus from the customer and the development team in order to support the decision process. That doesn't mean that tools like PowerPoint aren't useful for presenting information, but it should not be the architecture artifact tool of choice.

6.7.1 The Good and Bad Sides of Experience

Solid systems architectures are the heart of success for complex SoS. However, for the MDSE, there are other elements to program success that must be considered. Besides the Systems Architecture, the other major criteria the MDSE must embrace for successful program development, delivery, and operations are:

- **Management support for the program**: It is not only important that upper management support the program, it is vital that support is communicated throughout the program organization. If the systems/software/hardware/test personnel do not know whether the program is supported, their willingness to give their best for the program is suspect. Management must acknowledge, up front, that problems will arise and management must remain visibly behind the project at all times. It is common for managers to not understand the engineering and architectural design for the System of Systems program.
  The key for managers is to enlist the help of objective, skilled monitors and auditors to help in the assessment of program progress; not just cost and schedule, but technical progress and technical compliance as well. Many a manager has been fooled by relying only on cost and schedule as their only measures of system success, only to find out near the end that the system is not viable and/or usable. Also, be cautious of highly skilled systems/software engineers who believe they should be program leads. Often they have an agenda to "build the system the way they would like to," assuming they know more than the customer what should be built. Having skilled personnel to monitor the technical progress can help ensure that the requirements, goal, and objectives of the overall program are being met.
- **Sound systems/software engineering methodologies**: the methods, techniques, and constructs employed on the program must be only those required.
- **Solid technical leadership**: it is important to have Multidisciplinary Systems Engineers with experience working similar type programs:
- **Cost and Schedule profiles that are realizable**: ones that are implementable, given the requirements, goals, and constraints on the system. Many programs have failed do to unrealistic, shortened cost and schedule profiels.
- **Sensible engineering organizational structure**: one that maps into the functional aspects of the program, along with a personnel loading profile that matches the schedule profile.

While these are all necessary aspects of a successful program, they are not sufficient criteria to guarantee success of a program. There are, in fact, no criteria that can guarantee successful completion and delivery of a system. At the same time, failure is not necessarily a bad thing. Failure has intrinsic value in foreseeing and preventing future system failures and can be a great source of engineering judgment for the Multidisciplinary Systems Engineer. As System of Systems continues to increase in complexity as future systems employ Artificial Intelligence (AI) into their systems, it increases the difficulty of understand the systems; particularly their emergent behavior. One of the major difficulties with AI infused systems is development of System of Systems that produce results within acceptable confidence levels with predictability. These types of systems will require an entirely new set of methodologies for integration and testing, since they may have the capabilities to think, reason, and learn.

6.7.1.1 The Pitfalls of Resource Estimation

Estimating the resources required to develop and implement a given System of Systems design is fraught with perils the must be avoided by the Multidisciplinary Systems Engineer. Most groups go into the task of resource estimation with severe optimism about the amount of work an "average" engineer can perform (whether systems, software, hardware, test, maintenance, etc.). Realistic estimates are often thrown out for a variety of reasons, not the least of which is trying to undercut a competitor during the bidding process. Failure to proper estimate resources can decimate a project, driving major cost and schedule increases. However, failure to estimate the technical expertise required across the development effort will do more to derail a project than just underestimating cost and schedule. Beware of managers asking you:

> Can you give me a quick rough estimate, just to get us in the ball park? You won't be held to these numbers.

Be very wary, for you ***will*** be held accountable for this quick and rough estimate. Below are a handful of common pitfalls that must be avoided when performing resources allocation estimates for a given System of Systems architecture:

- **Padding**: often estimates are padded to account for a safety margin around the given estimate. Don't pad for pad's sake. Take the time to be certain of your estimates and risk assessments.
- **Optimism**: using all "optimistic" estimates in the resource planning stage. The engineering staff rarely works as efficiently as you would like to believe they can achieve.
- **Poor Risk Assessment**: again, the tendency is to be optimistic about the possible risks to the program. Do not assume everything will go according to plan, because it won't.
- **Missing Architecture Elements**: it's possible to miss something.

- **Unavailability of Required Resources**: before you assume the technical level of the staff you intend to use, make sure staff at those technical and maturity levels are actually available; verifying that the cost estimations are in line with the technical level you are assuming [87].
- **Underestimating the Project Scope**: if proper requirements decomposition/derivation is not performed, the overall project scope may not be understood, and the estimates will be wrong.

Remember, good estimates do not happen by luck, they happen by proper application of Multidisciplinary Systems Engineering practices and principles.

6.7.2 More Process Is Not the Answer

With failure, or at least major problems in a system development, there is a tendency to increase the amount of oversight or process within the Systems Engineering organization. Many programs have tried to resolve problems within the systems or software designs and implementations through substituting process, analysis, and "band aids" instead of taking the time to do the right engineering. The Multidisciplinary Systems Engineer should avoid this tendency. Without disciplined engineering and real creative processes, failure will continue, even if the increase in processes will make it take longer to realize continued failure is inevitable (many embrace this practice to fool themselves into thinking things are ok). The truth is process for process sake is never a solution. In general:

- Great processes, even when followed to the letter, cannot prevent failure if the engineering is not correct.
- Processes cannot replace management and/or leadership.
- Processes, even good processes, will never replace Systems Engineering. You cannot process your way to success.
- Processes will never fix a bad design.
- Processes cannot determine why a design is bad.

One of the significant problems with too much process is that is tends to remove accountability, critical thinking, and introspection. Managers tend to feel that if engineers would only follow the processes, things would be fine. And when they aren't, that means there aren't enough processes. This attitude and course of action never ends well.

6.8 Discussion

The architecture methods discussed here are but a few of the available choices, but are those that are seen as the most useful by many systems architects. As you gain more experience in MDSE you need to decide for yourselves which views you find most appropriate, which "frameworks" make the most sense, and which tools facilitate your overall System of Systems design. There is no one right answer, regardless of what many will try to tell you. In many cases, the views, the framework, and the tools, may be dictated by your customer. Therefore, it is incumbent on the MDSE to understand, at least at some level, all the available architecture and design methods.

Chapter 7
Systems Engineering Tasks and Products

7.1 Technical Planning

Developing of a cohesive technical plan, requires engineering efforts to follow established guidance at each tier of the architecture, and is critical to achieve successful system engineering processes. A technical plan should establish the guidance and/or work instructions for the technical artifacts, so that a standards-based approach can be maintained across both small and large development, taking into account time, schedule and personnel turn-over. When guidance is established, it allows for management and tracking of activities at each development level ensuring system design and implementation meets development objectives. As the program progresses, refinement of technical guidance may occur. If so, it is critical that changes in guidance flows down efficiently to the entire engineering staff and that the new direction is followed. Although many optimizations can be found during development, it is important for Systems Engineering to watch closely for and to avoid unnecessary shortcuts during all technical engineering efforts, as each shortcut can lead to increased development and cost risks.

A technical plan should establish iterative control gates or milestones. These control mechanisms manage and help optimize engineering activity and allow critical technical input, as well as, budget and schedule assessments to be performed at the appropriate times. Control gates should also include the established technical exit criteria used to assess quality and progress against the tasks within the plan. The exit criteria must be measurable and finite. Vague or subjective measurement for technical progress will almost certainly lead to a significant wave of technical deficiencies within the system engineering artifacts, which can cause exponential growth in additional activity over time. This obviously results in increased costs and schedule risks, and potential re-planning efforts or even loss of the program. To avoid a bow wave of unfinished technical objectives from occurring, a short work off period between the completions of a control-gate/milestone can be built into the technical plans for deficiency work-off, prior to the movement into the next phase

J.A. Crowder et al., *Multidisciplinary Systems Engineering*,
DOI 10.1007/978-3-319-22398-8_7

of the engineering/development. Additionally, as Systems Engineering iteratively pays close attention to the optimal size and discreteness of tasks, and builds that knowledge into the technical plan, then a program has potential to continuously build in program optimizations by design.

7.2 Technical Assessment

Technical assessment at key iteration points and/or milestones is important to ensure that the stakeholders and system engineers are able to incrementally validate progress against system objectives. The assessments, when done properly, allow the program to be assessed for technical quality, as well as progress relative to efficient development to schedule and budget. The technical assessments should include performance against the exit criteria for all tasks at the control gates and/or milestones, to include Technical Performance Measure (TPM) reports and actual measurements when available.

Formal reviews should be conducted at the control gates/milestones, where the measurable exit criteria, designs, TPMs and alike are assessed for satisfaction relative to the technical plan and guidance. As noted above, when deficiencies are encountered, the review may need to have a risk assessment performed to determine whether any greater underlying issues exist which could constitute redesign or other additional tasks to be performed to bring the design back to order. Progress past any milestone should only occur if the risk and deficiency work-off plan has been deemed to have balanced risk to be achieved during the work-off period, otherwise, it is important to keep the iteration/milestone event open and work-off the defects and/or redesign.

7.3 Technical Coordination

7.3.1 Key Decisions/Assumptions

Many times during development of a new or modified system, technical decisions and assumptions are made, but they are not documented properly. As the program progresses and changes, it is important for the engineering personnel to capture technical decisions and assumptions in a formal log, which should be available to other personnel. This log is maintained for the life of the development. These decisions and assumptions should be captured for all technical phases and artifacts. Numerous failed programs can be traced to continuous rework caused when technical decisions and assumptions are maintained the memory banks of the engineer/development staffs. The documentation of technical decisions/assumptions should be in a registry, to include decision makers, date and key decisions and assumptions. When older decisions or assumptions are changed, they should be retired but maintained for historical purposes.

7.3.2 *Change Control/Configuration Management*

In small, medium and large engineering efforts, it is very critical to maintain configuration management of engineering artifacts that have been approved for use during development. If changes are required during the development of a particular iteration, a formal change control must occur, so that the proposed changes specified by one area performing the development are properly communicated to the entire development team. The change control process is critical to technical quality, because, when it is not performed, the entire program may incur multiple rework cycles before the exit criteria is met for a given control gate and/or milestone. This rework almost always results in cost and schedule delays on the program.

7.3.3 *Technical Documentation/Engineering Database*

The CM process should provide engineers the capability to effectively manage the technical development, in formally controlled technical documents, engineering databases and development tools. The control system should also provide the capability to maintain different versions of the CM control materials, to allow baselines and exit criteria to be established and measured at the control gates. If engineering solutions do not include version controls, rework almost always occurs in the development efforts, as the various engineer/developers work to different sets of versioned artifacts. This is another lesson learned on many failed programs.

7.4 Business/Mission Analysis

An often overlooked effort within system engineering is the need to identify, collect and prioritize stakeholder needs. The system engineer should perform an initial mission assessment, and then continually refine the assessment, as needed, as the program progresses, and as new development changes are introduced by the stakeholders. This assessment should fully decompose the stakeholders new objectives, budget, schedule or system capabilities/requirements levied on the system. The analysis should also include a revisit of the key decisions and assumptions. When this has been performed, an injection point needs to be established in the engineering/development schedule for any new tasks. System engineering should also establish a new version of the baseline, and identify any impacted exit criteria for the follow-on tasks remaining in the development. When this does not occur, the risk of a successful technical solution increases.

7.5 Defining System Technical Requirements

Technical requirements are one of the most, if not the most, important aspects of system engineering. The task to build, manage and maintain them are not appreciated until satisfaction of stakeholder expectations fails, at which point, technical requirements become the most critical aspect of any discrepancy discussion.

While significant support data is included with requirements, such as use cases, architectural materials, design guides and alike, the technical requirements are the first and foremost SE artifact that must be verified and validated. Technical Requirements form the "contract" between system engineering and the stakeholders, as well as system engineering and the developers of the system's capabilities.

Definition: **Requirement**—A statement that mandates that "something" must be provided, produced, accomplished or changed.

Definition: **Requirement**—A statement that mandates "something" must be provided, produced, accomplished or changed.

Examples:

- The ground station shall collect solar flare metrics continuously.
- The diesel engineer shall maintain 535 hp at 2000 rpm at nominal cruising speed of 75 mph.

Before we discuss guidelines on the development of technical requirements, there are a number of topical areas that are addressed in the next few subsections to provide some corollary insights into proper requirement generation.

7.5.1 Formal vs. Informal Requirements

"*All systems have technical requirements.*" One can argue that this statement is not true; however, the distinction will be determined by comparison of formal verses informal control of the requirements. Formal controls are governed by processes and practices to ensure that the system being built, or procured, meets the end needs of the stakeholder. Informal requirements may or may not result in meeting the end needs of the stakeholders, and are measured using subjective methods of assessment, rather than distinct measurable criteria. Informal requirements typically have informal change control associated with them, forcing the developers or vendors to satisfy a moving target. Informal control can be good when building simpler, low-cost, non-complex systems like a sand castle on the beach, but do not work well in development activities when used to meet the end needs of a customer or management staff spending large sums of money on a large complex system.

All developments or procurements should recognize when informal requirements can and cannot be used. However, as a rule Systems Engineers must adeptly understand that outcomes of efforts employing informal requirements result in subjective interpretation and derivation and may be unreliable. This chapter is dedicated to technical requirements that are formally controlled, where the organization uses a formal tracking mechanism (e.g. Relational Database Management System (DBMs), managed word documents, spreadsheets and/or like materials that have versioning capabilities).

> ***Question to think about:***
> *What is the net effect of allowing formal requirements to age and become out of sync with the changes on the program?*

7.5.2 New Systems

Another aspect to technical requirements is associated with development of new systems verses upgrading and/or replacing existing systems. Since time eternal, we have developed "new" systems, but rarely are the systems new, rather they are new applications of thought and/or use new technology to solve existing problems. The SE must understand that a stakeholder, who wants a new system, must be prepared to accept a "new" development. This means that although the SE works with the stakeholder to determine customer needs, the system engineer is free to define the requirements to meet the objectives of the stakeholders (derived requirements), rather than working to a predefined requirement set (primary and/or contractual requirements) that has been updated from prior developments and/or procurements. Rarely is the system engineer allowed to define a new system purely from derived requirements outside of research and development, where new thought processes and applications of technology may be invented to solve the solution for a stakeholders'.

The point being made here is that most systems are enhancements to existing systems, and come with some predefined expectations (requirements) of what the stakeholders are looking for with the development and procurement activities. These existing requirements may have to be assessed for proper "Tiering" of the technical requirements set.

7.5.3 Acronym(s)

Today, we utilize acronyms as a form of communication. All acronyms used in technical requirements should be spelled out, and if there are multiple acronyms that are the same, but have different meaning (such as: ST = street, ST = state…), the acronym should be indexed (e.g. ST1 = street, ST2 = state…). Ensuring utilization of the correct acronym within each requirement is also very important, as they affect validation.

7.5.4 System Tiers

Another aspect of system engineering is the management of the requirements definition and quality within and across system tiers. All systems are comprised of architectural tiers, System tiers can be thought of as segmentation levels each of which represents a capability type. A well-known example of system tiers is represented top-down by: Visualization, Services, Applications and Data. Depending upon the level of fidelity desired, system tiers can be subdivided as needed.

Examples:

- Our "solar system" is comprised of a sun and nine (give or take a few) planets; therefore, you have a sun, and nine planets that comprise the "solar system".
- The governance of the United States of America, a very large system, is comprised of a federal government apparatus, 50 state governments and a few territories, each with their own levels of governance.
- A modern automobile is a system. It is comprised of multiple sub-systems, components and units (e.g. System = Automobile, Engine = Sub-system, Water pump = Component, Fan Wheel = unit).

The examples are meant to show that all systems have tiers. Systems engineers need to be aware that tiers affect the definition and the quality of technical requirements. Hence, the better the definition of the requirements and how tiers affect them, the more potential to effectively validate the requirements. Figure 7.1 illustrates a high-level view an "n-tier system architecture".

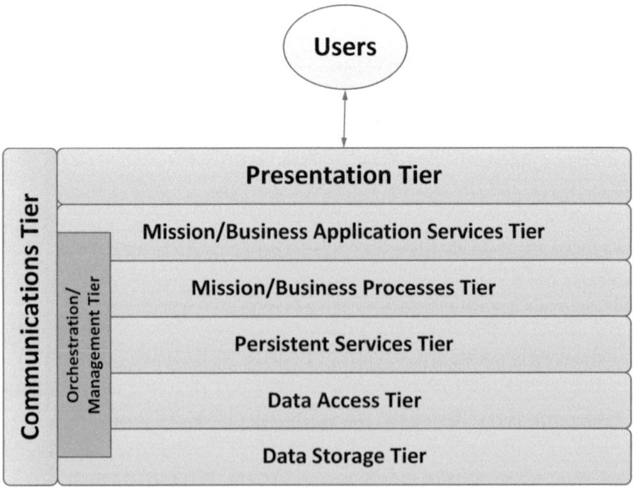

Fig. 7.1 High-level tiered architecture

7.5.4.1 Explanation of System Tiers

Presentation Tier: this tier responds to service requests from the Service Layer and issues service requests to the Mission/Business Process Tier. The Presentation Layer relieves the Service Layer of concern regarding syntactical differences in data representation within the end-user system.

Mission/Business Application Services Tier: this tier responds to service requests from the Presentation Tier and issues service requests to the Mission/Business Processes Tier. The Services Tier provides the mechanism for managing the dialogue between end-user business application processes. It provides and establishes check-pointing, adjournment, termination, and restart procedures.

Mission/Business Processes Tier: this provides services for an application program to ensure that effective communication with another application programs in the enterprise is possible. This processes tier is a service tier that provides the services necessary for the business applications to run and communicate effectively across the enterprise.

Persistent Services Tier: the tier allows application and service developers to work with a standard set of objects that read and save their state to a relational database or other storage formats (e.g., flat files). The Persistent Services Tier can be looked at as a way to access objects within the enterprise storage system. This includes data resource integration that abstracts out the information regarding data object storage such that the Persistent Services Tier contains the concept of Data Access Objects (DAOs). Data Access Objects are used by Java classes to handle all interaction with the Data Storage System. An API is provided so that other classes do not need to know how to create, retrieve, update, or delete a given object. Data Access Objects are defined by their Attributes, Object Key, and Events.

Data Access Tier: In the tiered architecture shown in Fig. 7.1, the need to persist data created at runtime is critical. The system's database clusters will normally use a relational database(s). The use of Data Access Objects for data access is indispensable in providing the methods to store database objects and flat files (e.g., log files and legacy data files from legacy systems).

Data Storage Tier: the use of Data Access Objects (DOAs) provide a unified, simple, and transparent persistence interface between the mission/business applications and data stores, and can significantly streamline the way the system deals with persistent data. The goal is to abstract away any persistence details and provide a clean, simple, object-oriented API to perform data storage. This drives another concept in Data Management called Data Labelling. Data Labelling is a way of controlling data on a row level. Each data row is given a label used to store information about data sensitivity. A label provides additional sophisticated access control rules in addition to those provided by privilege management. It further mediates access to a data row based on the identity and label of the user and the label of the row. This provides an additional level of privilege management to the overall system. Label-based privilege management is adjusted to conform with the systems privilege management policies dictates by the requirements, constraints, and operational concepts. Together these policies dictate the criteria by which access to an object is either permitted or denied.

Orchestration/Management Tier: this provides for automated arrangement, coordination, and/or management of complex services within the Mission/Business application layer and allows more direct access to data and data models for near-real time, computationally intensive Mission/Business applications. This may be done with or without human intervention.

Communications Tier: this tier (message handling) is an important part of the overall service-based system, in that it that provides for various types of messages and notifications for User-to-User, User-to-process, and process-to-process communications. It provides for both synchronous operations (generally for query) and asynchronous operations (generally for business transactions) within a single, unified framework.

Question to think about:
If you are going to try and build a "new system" using the "n-tier architecture" patters, what happens to the development if any of the tiers are removed from the design?

7.5.5 Categories of Technical Requirements

Another aspect for effective generation of technical requirements is working at understanding the categories of requirements that exist. An organization may or may not require categorization of requirements by type; however it is good practice to keep the requirement aligned by requirement type. As an example, as one is building a deliverable system, the functional requirements for the system should be aligned separately from the shipping requirements used for delivery.

In the case of the Department of Defense (DoD), governed by Federal Acquisition Regulations (FAR), the DoD requires a vendors meet numerous specifications and standards to be compliant with the FAR, in addition to the functional requirements assigned to the actual stakeholder needs. As such, technical requirements management should address all categories of requirements, because failure to comply with all requirements results in a discrepancy within the baseline, and comes with legal ramifications.

7.5.6 Primary/Contractual Requirements

A primary requirement typically defines a capability need within the system to meet the objectives the customer is trying to achieve. Primary requirements typically define who, what, when, where, how and constraints as needed to meet an objective.

Examples:

- The sales system shall provide automated sales summaries by close of business daily by employee, to maintain an office staff of 25 employees.
- The sales system shall provide automated (how) metrics summaries (what) by close of business daily (when) by employee (constraint), to maintain an office staff of 25 employees (objective).
- The sales system shall provide the capabilities for an office manager (who) to view the daily sales summaries by employee (what).

Included in primary (contractual) requirements provided to the system engineers, the stakeholders typically include learned lessons from other developments or procurements. These requirements may or may not, and probably are not, defined at the top tier of the system or objectives. This requires the systems engineers to properly tier the requirements in the decomposition of the system, and provide traceability to the primary requirement set [96]. This may require the systems engineer to derive system tier requirements, if the lower level tier requirement is not part of a higher level objective or primary requirement.

7.5.7 Derived Requirements

Derived requirements may occur as part of a new or modified procurement, and typically occur as part of the decomposition of the system from the Statement of Objectives (SoO) or primary requirement sets. Derived requirements form most all of the system requirements, unless the customer or management provided a completely decomposed system specification to the engineers. In most cases, even when decomposed requirements are provided, some derivation can still become necessary as the design is fleshed out over time [97].

Definition: Derived Requirement—a requirement that decomposes/refines the primary requirement or SoO.

Derived requirements allow the system engineers to elaborate with additional details in order to meet the Statement of Objectives (SoO) or contract. A derived requirement can impose specific solution sets to meet the end needs of the users.

7.5.8 Parent/Child Requirement Relationships

The primary requirements at the top tier of the system typically flow from a statement of objective or other primary definition. As you decompose the system, the next tiers of the derived requirements are children to the tier you derived them from. Each tier is decomposed from the prior tier, until you have reached the lowest level of definition needed by the developers or procurement vendor.

To ultimately satisfy the primary/contractual requirements in the system, the trace down to the lowest development tier needs to be maintained for validation of the primary requirements as you integrate and build up the system from the lowest tier to the deliverable system.

7.5.9 Technical Requirements and Attributes

A requirement needs a few key/recommended attributes:

- Unique ID—to distinguish it from all other requirements
- Requirement Text and/or Object Text—actual language defining technical need
- Allocation—aspect of the system the requirement belongs to (e.g., ground antenna, engine…)
- Validation Criteria—measurable criteria use to accept the capability works as specified
- Decomposition Trace—Parent Requirement ID

Once the key technical attributes are defined, each system engineering organization may have more required and optional attributes. There is no right and wrong number of attributes, but with each additional attribute added to a requirement structure, the more costly the requirement development and maintenance effort will become.

7.5.10 Language and Requirements

General Rule: Requirements must be unambiguous, clear and concise.

Language is naturally ambiguous. Even with well thought out objective text, and effort to achieve clear and concise language, the written outcome in the objective text will still undeniably come with large number of different interpretations of the requirement (e.g. "requirement lawyering"). Proper interpretation is based upon the requirements definition, usage, and system testing experience of the engineers reviewing or using the requirement for their efforts. To properly define a clear, concise, and unambiguous requirement, you may need to utilize additional characteristics to fully bound the contextual meaning of the objective text.

Because of the high potential of interpretations for any requirement, system engineering organization try to strike a balance between the language in the requirements and the number of attributes assigned and maintain in the requirements database. Once the stakeholders and developers interpretations of the requirements have been validated along with all of the attributes to be used, the system engineers should maintain the requirements and historical agreements made during requirement negotiations. If a requirement is being modified, the historical agreements should be maintained as well.

Proper maintenance on requirements becomes invaluable as the system progresses through the acquisition, development, delivery, and maintenance cycles on a system. The overall lifecycle on a system may range into decades on systems (e.g. Dams, Ships, Air Plane, building…), and the future engineers coming onto a project, need to be able to access the historical meaning applied to the requirements on the system, to avoid misdirecting future development initiatives.

7.5.11 Defining Technical Requirements

In the traditional structured system engineering design approaches, we allocate requirements by levels of the system tiers, by naming the tier (e.g. system, element, sub-element, sub-system…), which is found in numerous commercial and government procurements.

- System shall… or The system shall…
- Element shall… or The element shall…
- Subsystem shall… or The subsystem shall

The naming convention approach used in structured developments works well, to bound solutions to specific implementations desired, as well as bounding the solution space.

Question: "How does naming lead to bounding the solution space?"

Answer: Quite simply, the SE is acting like the software or hardware architect when making this allocation to specific portions of the system. Care should be used to ensure that the requirement applies to the portion of the system capabilities that the "shall" statements define.

As the system is decomposed through derived requirements, the derived technical requirements should pertain to one and only one element of the system. While the capabilities may be similar across elements or subsystems, they most certainly will be the same. As the branches of the system decompose from the top tier to lowest level units of software or hardware, those requirements should apply to one and only one branch of the decomposition. This ensures we have proper decomposition and avoids generic, open ended requirements in the lowest tiers of the architecture [98].

In another approach, we may have many specialty engineering teams involved with the development (DBM, HCI, RMA…), where we may want the "structured" data base team to do all DBM development for multiple portions of the system. As such, you should avoid writing requirement based on the specialty "team", and insure the requirements define the specialty "functionally" need by each element that is handed off to a specialty team to build.

With the move into more modern software and hardware approaches, such as Cloud and Service Oriented Architectures, the requirements take on new meanings, but can still utilize the structured approaches, which still leads to bounding solutions spaces offered by new applications and software constructs. Again, care needs to be used to avoid overly constraining the software and hardware architects [99].

7.6 Defining System Architecture and Development

Based upon stakeholder objectives and requirements, systems engineers must identify the major functional capabilities required in the system. By taking this first step, the functionality can then be assessed for use of existing system solutions, definition of new solutions, and identification of the hardware, software and operator procedures to be employed in the design. In large systems, the system may have multiple tiers of system structure (e.g. segments, elements, sub-elements and alike). When the system being developed has more than one functional tier, the system engineers must ensure that the functionality partitioned to each tier is properly isolated from the other capabilities in that tier in both requirements and architectural spaces. This is performed through functional decomposition and requirements allocation.

A good example of a very large system is satellite TV, where the system has a space segment and a ground segment. The capabilities applied to a satellite (element of the space segment) are unique in relation to the capabilities applied to the ground segment. Moving further down the decomposition path, within the ground segment there maybe one or more control stations (element), and one or more ground antennas (element). As you decompose the system capabilities along the lines of the control station, you will eventually have computers, networks and alike. This will be true for the ground antennas. The decomposition of both functional capabilities and requirements from segment to the Line Replaceable Unit (LRU) of an element must flow down and be unique to each element.

Also, while you may have specialized engineering team develop solutions for say the networks on both the control station and the ground antenna; the requirements and architectural component for each element should be uniquely defined for one and only one element. This allows each element to be sold off individually. This is the difference between system engineering and development. Systems Engineering is responsible for the proper decomposition of each element in the system; whereas, the development teams may utilize specialized developers to build like components for different elements.

Systems Engineers must avoid shortcuts in functional decomposition and synthesis. The requirements for a control station are not the same as the requirement for the ground antenna. A good number of systems engineers get trapped into implementation space, which is bad practice. Just because the COTS H/W or S/W is the same, the COTS H/W and S/W are still unique in systems engineering spaces between functional elements. A pair of computers load balanced on the control stations may be exactly the same as the single computer on the GA, but the only similarity is the acquisition specification. To be proper configured for the two different elements, they should have different requirements and verifications methods in relation to the functions they are supporting [100]. Very rarely do the requirements equal in relation to the functions on the different element; otherwise, you have incorrectly decomposed the functional capabilities and element assignments.

Once the functionality of a tier is defined and isolated relative to each other, the system engineer must partition the capabilities into hardware, software, and procedural components. Partitioning capabilities to hardware comes with the need to evaluate commercial-off-the-shelf capability verses custom development for specialized or unique hardware to satisfy the functionality. Over the past few decades, development of specialized hardware has slowly been replaced by the commercial market place. There are still unique hardware developments, but more and more options exist in the COTS market place.

Software partitioning also has benefited from the COTS market place as well, where numerous functions can be procured rather than custom coded. A word of warning, the more COTS or FOSS you select for use in the software functional solutions, the higher the integration costs and the more risk you incurred, which is caused by software interface code, secure coding standards and alike. Also, since the market place continually moves forward, the ability to maintain a solution with high numbers of COTS/FOSS integrated together becomes less feasible in relation to the system lifecycle. It may look great in the proposal, but in practice is becomes a nightmare for the integration, test as well as maintenance organizations, as the commercial market place leaves the system aging rapidly.

Lastly, when allocating functional requirements to operational procedures, the operational procedures should be assigned to the operator only when the operator is required to make management decisions or provide necessary inputs to the system. The operator should not be the required capability in the system to make the hardware and software work properly. Another words, the operator should not be in place to execute standard and non-standard workarounds to system failures.

The allocation of functionality to hardware, software and/or operational procedures must then be evaluated and assessed to see if alternative solutions may provide improvement in the design relative to the stakeholder's objectives. As an example, a fully unmanned system may eliminate operator errors; however, unmanned systems typically require completely fault tolerant performance and high reliability, leading to increased development costs [64].

The alternative solutions may be as simple as a make/buy assessment, to a fully executed trade study. In the worst case scenarios, multiple prototypes of alternative solutions may be required, where the development of the prototypes, process and practices employed in the development is fed into a trade study assessment. Trade studies are covered later in this chapter.

Once the higher level architecture operations have been established, the evaluation process leads to the selection of the preferred system architectural solution. The selection process should include the stakeholder objectives, the each alterative system solution, and selection criteria. Each alterative solution is weighed or scored in relation to the selection criteria, balanced against the objectives and requirements satisfaction. Once the stakeholder, engineers and developers have made the selection, you are not off and running to the next tier of the architecture.

While many different organization go about selling "their" architecture, most information systems with user/operators, start to look very similar except in the terminology used to describe them.

7.7 Further Decomposing the System

Once the high level system architecture is established, and continuing with the satellite TV example with focus on the control center (sub-element) within the ground station (element), and using the "N-tier" architecture, the requirements defining the control center provide for the processing areas of the architecture.

In the older structured designs methodologies, they would be called sub-systems, in the modern services approach they would be called service domains. The requirements allocated to the ground station/control center must be decomposed into the sub-system and their functionality. As such, the "N-Tier" architecture provides for a predefined solution set for operator controlled processing centers, where the presentation tier would be considered the User Display Subsystem or User Display Services as noted in the prior section. The element requirements are decomposed into the various types of display needed by the operators and maintainers of the system at the subsystem tier, which eventually will be decomposed down the components of the displays. This decomposition allows for the system engineers to evaluate COTS products, or submit the requirements to development team, and vendors in order to support make/buy and/or trade study evaluations.

As discussed in other sections of this chapter, the development teams may be aligned based on specialties, and in this example, the User Display Services may be built by a team of Human-System Interface specialists. They should be used to build all of the displays in the system, or provide oversight to vendors selected in a make/buy decision. This ensures a common look and feel across elements with different display requirements, and reduces the risk of divergent or integration problems with the display framework.

As discussed in the requirement section of this chapter, the overall purpose of the allocation of system requirements and decomposition of the requirements from segment to component and/or unit of hardware/software/operational procedure provides the capability to ensure completeness of the system and architecture in relation to stakeholder needs and system requirements. Once the decomposition is done, the lower level specification must follow the same rigor as higher level tiers, and be allocated to the architectural components and units of the sub-systems for the element to be developed. The lower level specifications should be documented and controlled just like all other requirements in the system.

7.8 Determining External and Internal Interfaces

In most cases, the stakeholders of a project typically understand what interfaces from external entities are required. But when the projects are exploratory, the system engineers will have to evaluate the objection and requirements, to identify sources of data and interfaces that may be external to the organization associated with the project. In both case, system engineers still need to look at all interfaces and ensure that all sources of data are address in external interface definition [101].

When a new development is replacing an older system, there may be many temporary data sources required to ensure all of the data needed for transition of the capabilities occurs smoothly. Systems engineering should provide contingency time and or risk mitigation plans in place, so that last minute changes for temporary interfaces can be incorporated and still meet the technical and schedule objectives.

Internal interfaces for the system being developed, typically are managed the same for intra and inter element data flows. However, Inter-Element interfaces typically will have more security applied to them, than the Intra-Element interfaces. At each tier of the system, the internal system interfaces have to be defined and documented, allowing for increasing levels of fidelity until you reach the build tier of the system. This allows the architecture model to be document and validated at each tier of the architecture. When this does not occur, the development will always incur rework, to include higher integration risks, as well as higher costs and schedule delays. Interfaces both external and internal must be validated in relation to Server and Client sides of the interface. This ensures that the Server and Clients are all in synch relative to interface responsibilities. This avoids the 4 pin connector, plugging into a 3 pin slot types of deficiencies.

Lastly, all interfaces must account for all modes of operations. When only one mode of operations is evaluated, the happy path is usually the only one evaluated. All modes of operations must be addressed, and accounted for in the interface design. A good example is test mode. If the interfaces do not provide for test mode/test flags to be applied to the data, the internal and external systems must be manually operated and or isolated from the system under test [102]. Modes of operations should address off-line as well as on-line modes of operations, as well as normal, test, and contingency operations. Systems engineers should always plan for the absolute worst case mode of operation when defining interfaces. This ensures that both clients and servers do not have to enter varying forms of contingency recovery or contingency operations when something within the system becomes defective. While they may be in a contingency recovery/operational situation, the system gracefully enters and exits these situations when the planning and engineering has accounted for it in the system design [103].

To avoid problems with external interfaces, a good practice in systems engineering is to design a decoupling mechanism between the external systems/interfaces and the system under development. Business-to-Business (B2B) concepts allow for changes on both sides of the interfaces to be decoupled from each other. This provides a focal point between internal change and external change to be isolated to a single change point, rather than rippling both internal and external systems. In today's market place, XML and "optional" tags on data is an example of decoupling data. If one side of the interface is not ready for a change, the data can be tagged as optional, and if the service agreement on the interface has been defined properly, the receiving system will ignore "optional" field until they are ready to use them. Hopefully, as systems engineer matures, both internal and external interfaces will focus more on management of change for an interface rather than just ports/protocols/data types.

7.9 Trade Studies

The purpose of a Trade Study is to support the technical needs of the Systems Engineering process throughout the development lifecycle. It allows Systems Engineering to evaluate alternative solutions, balancing cost, performance, compatibility, testability, supportability, etc., in light of the systems requirements, functionality, quality attributes, configurations, constraints, and CONOPS. Rarely is there only one way to solve an issue in the Systems Engineering design process. Alternatives must be evaluated and traded against criteria, each weighted as to its importance to the overall system, to arrive at a balanced design in terms of overall cost, schedule, and performance for the system. A Trade Study can help Systems Engineering develop and/or refine a System Architecture concept and determine if additional analysis, prototyping (synthesis) or other Trade Studies are required to make a design decision. One classic example of a Trade Study that is performed as a normal part of System Design is known as the "Build vs. Buy" Trade Study [104]. This Trade Study is used to determine whether it is advantageous to build a given set of capability as part of the system development, or to buy a Commercial-off-the-Shelf solutions that is built by a third party vendor. The major steps in the Trade Study process our illustrated in Fig. 7.2.

7.10 Life Cycle Cost Modeling

One of the major drivers factored into the overall decision to build any given system is cost. Both the initial development cost of development, implementation, test and delivery and the cost of operations and maintenance over the lifetime of the

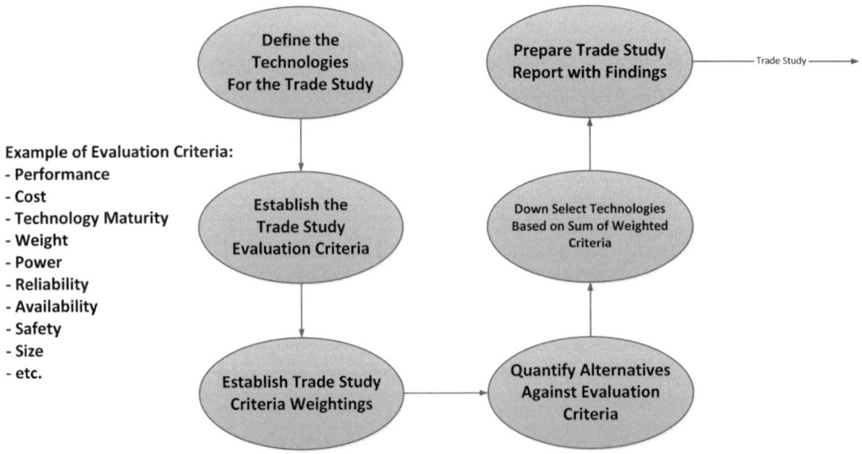

Fig. 7.2 The overall trade study process

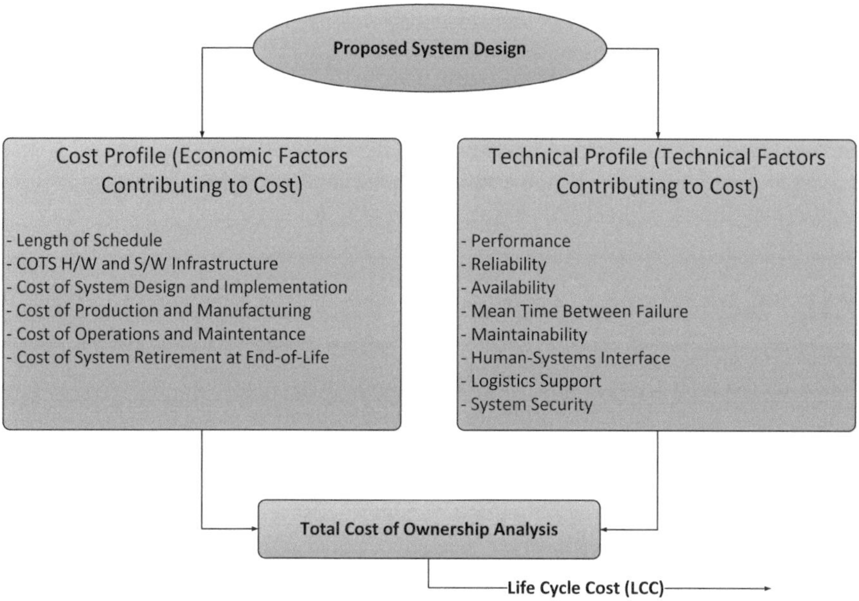

Fig. 7.3 Factors contributing to lifecycle cost analysis

operational system are important cost drivers. Together, they form the overall Life Cycle Cost (LCC) for the system. One of the roles Systems Engineering plays is to help assess creating the Development and the Operations and Support (O&S) models required to assess the overall costs of the system, the Total Cost of Ownership (TCO) to present to the customer. Within the Lifecycle Cost Modeling analysis, both cost drivers and technical drivers must be traded off in order to arrive at a Total Cost of Ownership that still meets both the cost profile (total cost spread across the schedule, taking into account year-by-year spending predictions) and technical profile (technical factors, like performance, that drive the design cost). Figure 7.3 illustrates this.

- The Lifecycle Cost Analysis helps to ensure that the O&S costs associated with operational readiness are considered in the overall Total Cost of Ownership decisions. Once the contributing factors to lifecycle costs have been determined, a modeling tool is utilized to estimate the TCO. One such modeling tool is the Lifecycle Modeling Language, which is an open standard language designed for use the System Engineering organizations to support estimation of lifecycle costs through the conceptual, implementation, installation, operations, support, and system retirement phases of the proposed system design. The Lifecycle Modeling Language was created to provide a common modeling language to trade proposed architecture and system design alternatives [105]. The LML provides simple communications of cost, schedule, and performance across each of the alternatives, creating a logical ontology of all information; information provided and information computed by the models. LML has six main goals:

- Ease of understanding by Systems Engineering
- Ease of understanding by customers
- Ease of extending the modeling constructs
- Support functional and object-oriented design approaches
- Support the entire system life cycle, from procurement to retirement
- Support evolutionary and revolutionary proposed system changes over the lifetime of each of the proposed system solutions

7.11 Technical Risk Analysis

Since all projects carry some elements of risk, it is important for the Systems Engineers to help define the program risks and mitigation plans to manage and/or reduce the identified risks. Risk reduction is aided by proper decomposition of the system, analyzing each subsystem, enterprise service, and/or component. Formally, a risk is defined as the possibility that a technical, cost or schedule target is not met because something planned does not happen, or because something unexpected does happen. Risk Analysis is the processes of identifying events that pose possible risks to the program, assessing the potential for the risk events, called Likelihood of Occurrence (LoO), and the impacts on the program, should the risk events happen. Once the risks and their LoOs and impacts have been assessed, plans must be put in place to minimize their effects on the program, either by reducing the LoO, or by reducing the overall impact to the program, should the risk events occur. It is important to understand that some form of Risk Management is essential, even for small projects. Without formal risk management within a program, the program is likely to:

- Assign inadequate resources for risk management
- Make decisions without knowledge of potential risks
- Repeat mistakes of the past
- Fail to execute the program on time, on budget, or not meeting technical requirements because of risk events.

Technical Risk Analysis and Management include (see Fig. 7.4):

- Planning for risk at the beginning and throughout the lifecycle of the program
- Continually assessing the program for potential risks
- Assessing identified risk events
- Developing risk event handling procedures/options
- Monitoring risks to determine how the LoO and impacts have changed
- Documenting both the process and the outcomes.

Overall system success and effectiveness is maintained throughout the development process by continuous and early detection/prediction of risk events and predictive mitigation plans to reduce degradation of Cost, Schedule, or Technical performance. Without the risk analysis and prediction cycles, the program is left to react to events as they occur. Figure 7.5 illustrates the potential for performance degradation, given predictive vs. reactionary risk identification and mitigation.

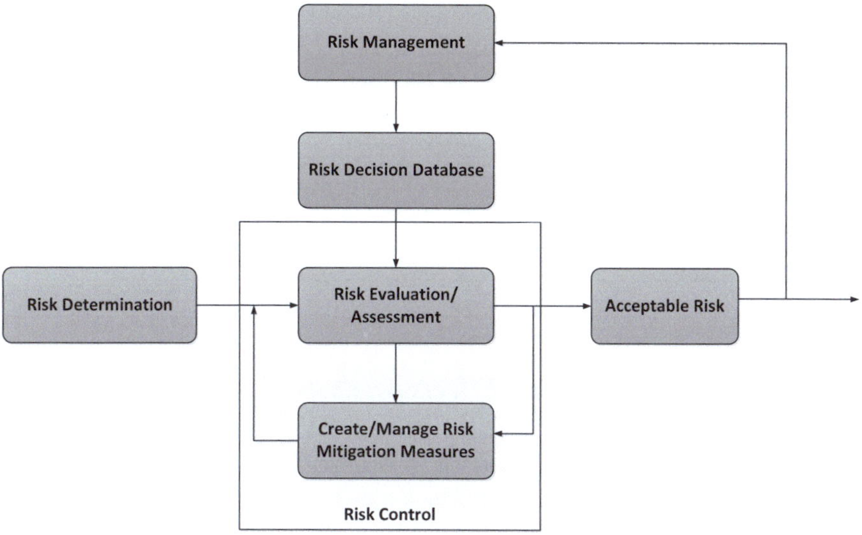

Fig. 7.4 High-level risk analysis and management process

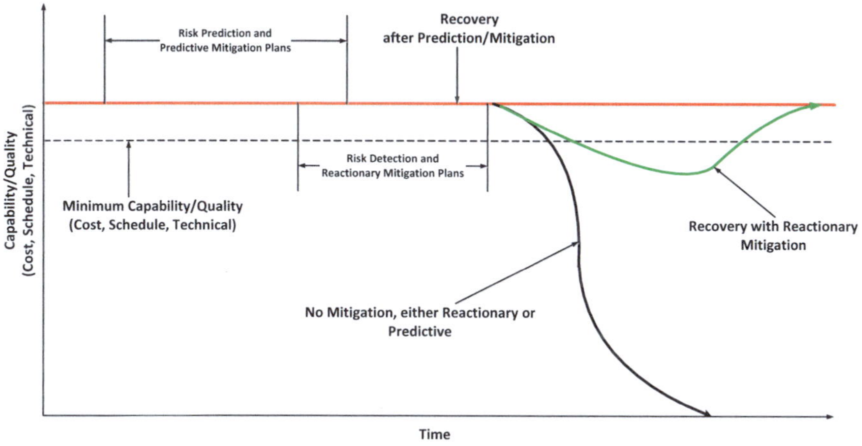

Fig. 7.5 Risk mitigation timeline: predictive mitigation vs. reactionary mitigation

7.12 Safety and Quality

All systems have to provide for safety of personnel, and in a lot of cases, safety of the system and its features. Therefore, Systems Engineers are tasked to evaluate all features of the system, and identify safety hazards. These hazards vary based on the actually hardware, software and procedures employed in the architecture. Some of the safety features that must be identified are electrical hazards to personnel and equipment/facilities, electrical and/or radio frequency emissions, sharp objects, weights of racks/components, movement hazards, etc. These identified hazards should result in requirements that can be allocated to the various components of the architecture for inclusion in the design of hardware, software and operator procedural solutions as applicable to the system design.

Once all of the safety features are applied to the architecture, the architectural solution must be assessed for quality and dependability in relation to the objectives and requirements. This is typically performed by processes such as Failure Modes and Effects Analysis.

7.12.1 Failure Modes and Effects Analysis (FMEA)

Failure Modes and Effects Analysis (FMEA) is an engineering tool utilized to assess the risks associated with the ways in which a given system can fail (called failure modes). The purpose is to identify these system failures, the conditions under which the failures occur, the effects these failures have on the system, and an analysis of how they system might be redesigned to eliminate or reduce the identified potential failure modes. The FMEA is predominately an inductive logical process which seeks discover what would happen if each of the failures occurred. The FMEA analysis process ties into the Risk Analysis process discussed in Sect. 7.11 to provide a quantitative analysis of the risks associated with each of the failure modes (or events); aiding in comparing competing designs. The high-level FMEA process is illustrated in Fig. 7.6.

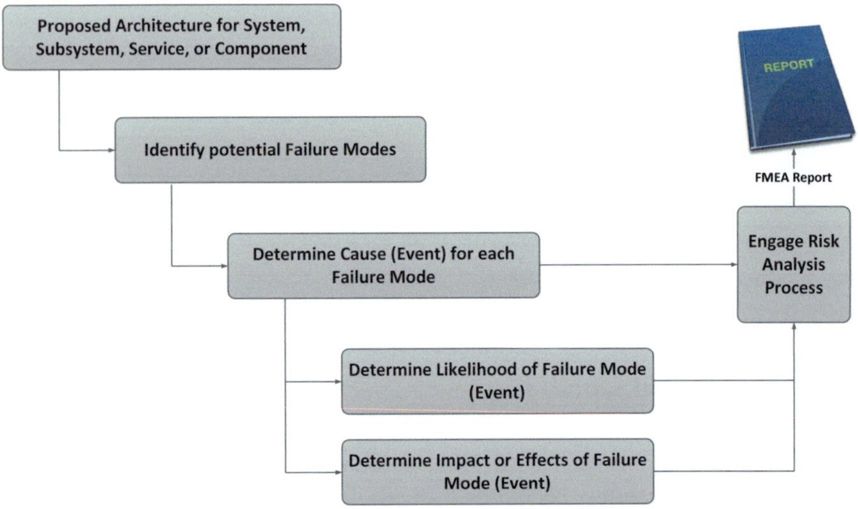

Fig. 7.6 The basic FMEA process

7.13 Logistics Support

The selection of hardware, software, operational procedures, with the inclusion of the FMEA drives the logistics support requirements on the system. This includes sparing for hardware, licensing for software, and safety procedures to be included in the complete set of procedures applied to the human in the loop. In many cases, the logistics support requires formal configuration management to be applied in the support environment to the operational system. Because of this the logistics requirements and supporting CM requirements should be treated with the same rigor as the rest of the requirements sets and architectural components. The sell-off logistics support requirements will vary based on whether it's an inspection requirement, or a functional test requirement. Sparing may have been computed on simple formula such as 10 % or 2 deep, where the verification is provided when the correct quantity of LRU spares are provided to the maintenance depot; whereas, the ability to build and delivery the software baseline to one or more elements, would be tested with the same rigor as functional capabilities within the system.

Logistics for all systems leads to the success or failure of the development. Only when the system being developed is a "throw-away" solution, does logistics not count, but even then, there may be metric collection during the short life expectancy that must be collected for future developments.

Logistics should typically be stood up in advance of the actual deployment of the system. This gives the maintenance personnel time to startup their organizations and supply chains, complete training and like. A system should never go into service without formal maintenance teams in place, provided by either the developers or independent organizations.

7.14 Verification and Validation

The right side of Fig. 1.8 (shown below as Fig. 7.7) illustrates the V&V side of system engineering. All requirements for hardware, software and procedures should come with verification and validation methods. If they don't, they are not requirements and should be removed from the system. Since most requirements are vetted by the stakeholders or management, this should never happen.

- Test—is the practice of conducting specific series of test with specific inputs and expected output for the capability. Most testing should be automated, to allow for repetitive testing during system build up.
- Demonstration—similar to test, demonstration is more fluid, and typically is validated by expected behaviors based on scenarios driving the system, rather than fixed input/outcomes of test. Typically, demonstrations are time consuming and go through a series of pre-demonstration dry runs to ensure the demonstration runs smoothly.

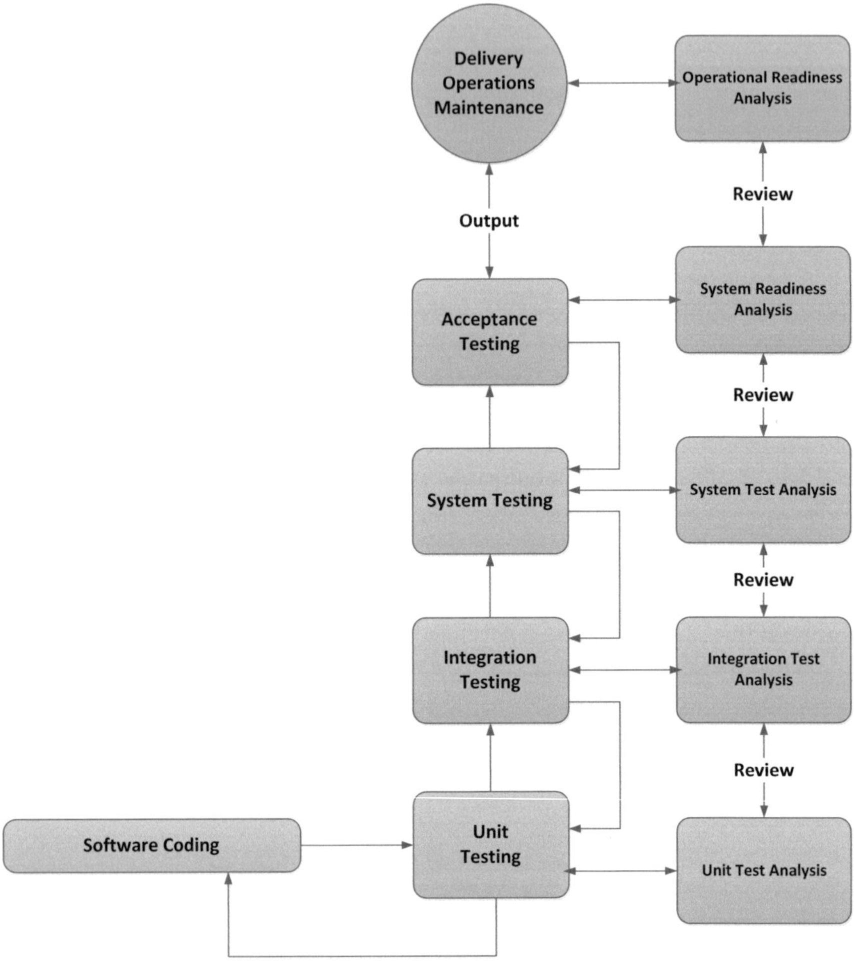

Fig. 7.7 The V&V process

- Analysis is used to validate a capability by the use of metrics or post-test/demonstration output data products, such as the evaluation of the collection of throughput times on an interface, where the statistical average is computed for nominal performance.
- Inspection is used to validate requirements criteria by using physical evaluation of the product, such as sharp edges, software code vice the PPS&C, etc. A good example of inspection is where the capability must be UL or FCC certified. In this case, the inspection is a simple validation of the certification supplied with the equipment, and the label on the equipment showing the UL or FCC label.

Each requirement should have one or more methods selected for V&V, and the selection will depend on the type of capability. A requirement that is automatically tested with each build, may never have to have secondary methods. If a capability requires extensive demonstrations on the initial test cycle, and receives updates on the next release, the new capabilities in the new release should be fully demonstrated, whereas, the older capabilities may incur a selected spot check/regression demonstration to ensure breakages between releases does not occur. Typically, Analysis and Inspection are performed together, but there are cases where some capabilities do not require both. Inspecting corners on an equipment rack will not needs an analysis, either it meets the acceptance criteria, or its. In contrast, a manufacturing inspection on screws may sample 100 of the screws every 10,000 screws, and then by analysis of the inspection, outcomes on the samples, the entire lots may be accepted. Any time analysis is used, statistical methods must be apply to the requirements acceptance criteria, and should include contingency plans to address variances for out of tolerance assessments. It would be madness for one million screws to be physically inspected.

Like the system architecture, planning for verification and validation is critical to system engineer. If insufficient/incorrect test methods are applied to the requirements, or insufficient planning for the acceptance program, the projects will always fail to delivery on time fully validated. The test program should have very detailed testing at the unit tier (hopefully you'll employ automation), where component integration may being the process of sample checking on capabilities, leading the system tier where demonstration of key events are performed. During the V&V program, the demonstration procedures should be validated relative to the expected system performance. When data is collected via test, inspection or analysis, the data from these events should be evaluated for validity, and then for verification and validation of the capabilities.

The final testing phase of the program is the most important, as most of the capital has been expended prior to final delivery. This phase, typically called "System Transition" is far too often neglected, and results in failed stakeholder objectives. Typically, validation of the capabilities is performed during the final system test phase, typically in the users' facilities, where the users of the system conduct operational evaluation of the verified system. The build up to final system test includes all internal, external interfaces that could not be validated in the development's test labs, or where the capabilities have to be connected to other parts of the system in the final configuration.

7.15 Establishing and Controlling the Technical Baseline

Once the requirements, architecture, and validation program is established, the baseline must be formally controlled. By formally controlling the baseline, the systems engineers are able to proper assess impacts of changes driven by the stakeholders, changes in the development its self, of changed caused by changes in the market place. If the system is developed in multiple increments, the formal baseline must be "captured" for the increment that is about to be started. Meaning, the system engineers must establish an incremental baseline for the iteration, which includes the partial set of requirements, architecture and testing that is to be performed to satisfy the iteration as defined in Sect. 7.1 of this chapter.

It is highly encouraged to avoid changing requirements, architecture and testing procedures during the test iterations. This requires the systems engineers to evaluate the proposed changes on the overall set of artifacts, and determine if an on-ramp can be achieved during the iteration. If not, the proposed changes need to be factored into future iterations. If the baseline is a tradition waterfall development, the program must account for on-ramp periods throughout the development cycle. This then starts to resemble iterative developments. Therefore, it is also highly encouraged to plan all development along iterative lines, unless the development is short fused and will not incur change.

7.16 Case Study: Why Good Requirements Are Crucial—The Denver International Airport Baggage Handling System

According to the Software Engineering Institute (SEI), the top two factors that result project failure are inadequate requirement specifications, and changes in requirements throughout the development cycle.

Denver International Airport: The Denver International Airport (DIA) is the largest airport in the United States (area), with the longest public runway in the United States. During the original construction of DIA a new, state-of-the-art baggage system was supposed to be built that was touted as the most advanced baggage handling system in the world. What it is now touted as is one of the clearest examples of project failure in the last 10–15 years. The baggage system that was originally proposed was supposed to be designed to automate the entire baggage handling activities across all three airport concourses for both outgoing and incoming bags.

One major problem was the failure to properly understand and decompose the high-level requirements and understand the overall complexity of the system being proposed. In addition, the design team did not adequately assess the risks involved in the new system. As a result of improper requirements derivation/decomposition, there were major problems building the baggage handling system and these issues caused a 16 month delay in the opening of the airport, while the engineering team tried to salvage the system and find something that would work and was acceptable. The delay in construction and testing of the baggage handling system resulted in a $560 million dollar (US dollars) overrun, and was featured in many articles as a major failure of the new airport.

The final design solution that was implemented at DIA was far less capable than was originally envisioned, providing automated only the outbound baggage of one of DIA's three concourses. The rest of the baggage is handled with a manual system. Even the one concourse's automated baggage handling system had to be scrapped completely in 2005 due to improper functionality. Having not been designed properly for maintenance, the $1 million monthly maintenance costs proved too expensive and going to a completely manual system reduced the overall lifecycle costs. In general the major issues with the baggage system were:

- The complexity of the System of Systems architecture
- Changes in system requirements
- Underestimating cost and schedule
- Improper Risk Analysis
- Failure to design and factor in backup and recovery systems

7.17 Discussion

The life of a MDSE is many things, as we have discussed throughout the book. Most reviews and discussions of Systems Engineering emphasize all of the other things other than requirements that Systems Engineers are responsible for. And while that is true, the one thing all Systems Engineers MUST get right, and the one thing that is at the heart of effective MDSE is getting the requirements right! Failure to properly deal with requirements, failure to do a proper decomposition, derivation, and allocation of requirements results in an almost impossible task of delivering an operationally viable system. Requirement decomposition is the heart of overall design quality. There are two major measures of good requirements decomposition, requirements coupling (linking) and requirements cohesion (functional correlation).

7.17.1 Requirements Linking

Linking describes the relative independence among requirements at each level of the requirements decomposition. Requirements linking measures how requirements and requirement modules (e.g., subsystems, CIs, etc.) are connected to other modules within the architecture. One of the factors that affect linking is the interface complexity between modules. The MDSE must strive for the lowest possible coupling between modules.

7.17.2 Functional Correlation

Functional correlation measures the specialization of modules within the System of System design. This must be measured throughout the levels of the System of Systems design, Element, Subsystem, CI, Component, etc. Ideally, each module

Table 7.1 Types of functional correlation

Correlation type	Quality of the module	Description
Coincidental	Disastrous	Module functions have no correlation, resulting in improper decomposition
Logical	Terrible	Multiple module functions that are conceptually related, resulting in code that is often overly complex with flags used to control the logical flow
Temporal	Very poor	Module written for many tasks that happen at the same time, resulting in the module being very difficult to reuse
Procedural	Good	Module tasks are related by the order of execution in that control flows from one to the next, resulting in code that is difficult to maintain
Communicational	Good	Module tasks use the same data, resulting in code that is a difficult to maintain, especially if the data changes for one task but not the other
Sequential	Good	Module tasks are such that the output of one feeds the next task, resulting in lack of reusability if only one task is to be reused
Full functional correlation	Excellent	All module tasks contribute to a single, related problem, resulting in modules that are inherently reusable

within the overall architecture has a specific function and is not written so that it accomplishes many uncorrelated functions within the same module. Low Functional Correlation results in having a module (e.g., CI) that does many uncorrelated functions. Table 7.1 illustrates the types of Functional Correlation and the results:

It is not enough to architect at high-levels, architecture should drive throughout each tier of the architecture and provide guidance throughout. The MDSE team(s) should be involved in every aspect of the system.

Chapter 8
Multidisciplinary Systems Engineering Processes

Multidisciplinary System Engineering provides the processes and oversight for the three main critical aspects of System of Systems design:

1. System Design and Development
2. Site Installation and Checkout
3. System Transition to Operations

The basic System Engineering Process diagram (Fig. 1.6) illustrated the flow of requirements traceability throughout a program's lifecycle and the basic Software Engineering Process diagram (Fig. 1.7) depicts the flow of software development from requirements through delivery of the system to the intended operational site [106]. We will now combine these to explain the entire System of Systems development lifecycle, which includes the decomposition/derivation flow of requirements, and provides traceability all the way down to the development items at the development level [107]. At each level the Multidisciplinary Systems Engineering organization and the customer have full visibility into the traceable matrices for all requirements. The Multidisciplinary System Engineering team works closely with the Systems Integration (SI) team in the development and integration of the system into the operational environment. The Multidisciplinary System Engineering process is integrated into the overall System of Systems Lifecycle Framework, as illustrated in Fig. 8.1. The Systems Engineering process depicted is one of the primary influences on Multidisciplinary System Engineering and on the development of the System Engineering plans required for programs.

Figure 8.1 provides a guideline to defining the System of Systems Engineering approach; providing a disciplined System of Systems Engineering processes that establish the framework for program design, development, production, and deployment. These processes utilize an architectural framework and incorporate key tenets of the overall Capability Maturity Model Integration (CMMI) process model.

J.A. Crowder et al., *Multidisciplinary Systems Engineering*,
DOI 10.1007/978-3-319-22398-8_8

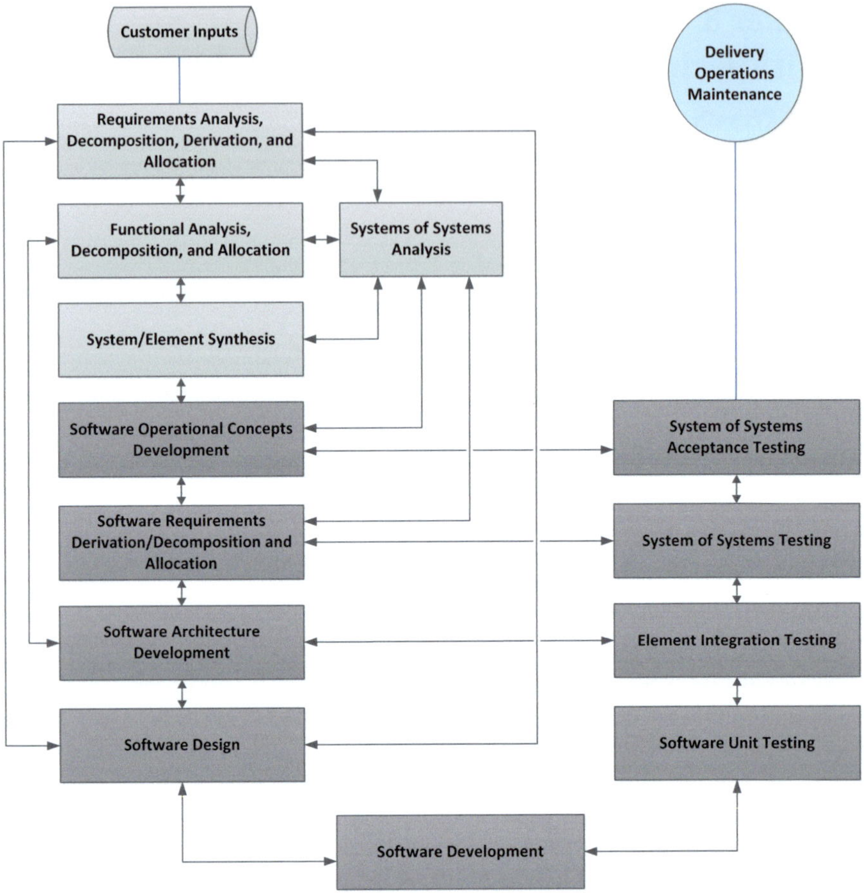

Fig. 8.1 Multidisciplinary systems engineering systems of systems process flow

Implementation of these processes can help in development of a fully capable system delivered within all technical and programmatic constraints. Not all processes are required for all systems. The Chief Multidisciplinary Systems Engineer must decide which processes are required, given the requirements and scope of the System of Systems to be developed. In this chapter, we will define program structures that provide the optimal planning and execution framework for all phases of program life cycle. We define high-level program structures that can be used to help define the overall System of Systems.

The System Breakdown Structure (SBS), the Work Breakdown Structure (WBS), and the Organizational Breakdown Structure (OBS), are three "pillars" of program structure which, when fully aligned, provide a full understanding of the Multidisciplinary System Engineering technical development.

8.1 High-Level Program Structures

8.1.1 System Breakdown Structure (SBS)

The System Breakdown Structure (SBS) defines the System of Systems elements and their related life cycle processes. The SBS also includes the definition of the program technical reviews and organizational structure required to accomplish the requirements, goals, and objectives of the System of Systems program. Figure 8.2 illustrates a high-level SBS for a desktop computer system.

8.1.2 Work Breakdown Structure (WBS)

The Work Breakdown Structure (WBS) is a decomposition of the System Breakdown Structure (SBS) that defines (at a high-level) all of the work categories required for development of the System of Systems the Multidisciplinary Systems Engineer must architect, design, implement, integrate, test, and deliver. The hierarchical breakdown that the WBS provides illustrates a clear, logical grouping of functions, or activities. This decomposition can be created in a variety of methodologies:

- Deliverable-centric
- Subsystems/components-centric
- Process-centric

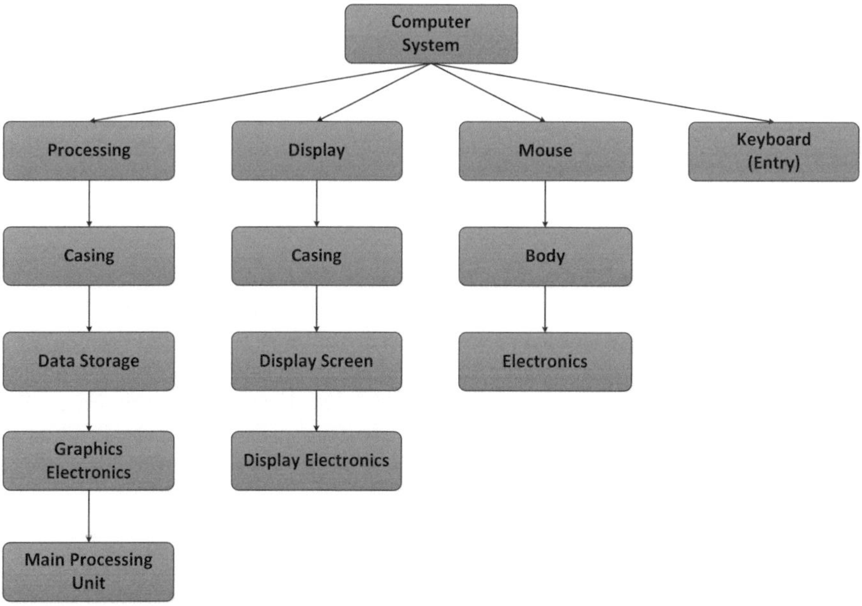

Fig. 8.2 Basic system breakdown structure

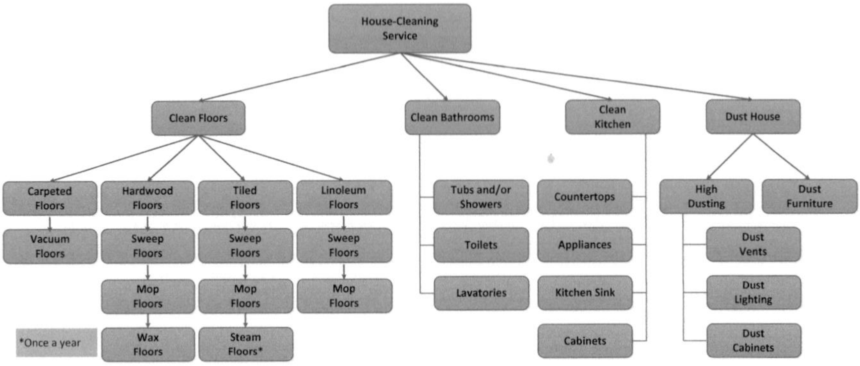

Fig. 8.3 WBS example: House Cleaning Service

All of the entries in the WBS should be reflected in other program documentation, e.g., requirements, objectives, program scope definition, at a minimum. The WBS allows a schedule to be created for the design, implementation, and testing of the system and allows the overall System of Systems to be broken down into manageable divisions. Typically a Program Management tool used to create a program plan and program schedule, the WBS can be an invaluable tool to the Multidisciplinary Systems Engineer (MDSE) that allows him/her to play the Systems Architecture against the WBS to see if the architecture supports the hierarchical WBS breakdown [19]. They are not identical structures, but they will aid the MDSE in ensuring the architecture is realizable and aids in identifying the deliverables at every phase of the project. A WBS can be created for any type of program or project and even simplistic projects can benefit from a WBS; it aids in thinking through how to organize and implement all of the pieces or tasks required. Figure 8.3 illustrates a basic WBS, this one for a house-cleaning service.

8.1.3 Organizational Breakdown Structure (OBS)

The Organizational Breakdown Structure (OBS) illustrates how (personnel-wise) the program will be organized into development or process teams. The OBS is, perhaps, the easiest of the design artifacts to understand, yet it communicates the development/implementation team structure and dynamics easily and helps establish reporting responsibilities across the System of Systems development program. Figure 8.4 provides a high-level illustration of an OBS. The OBS is created in a typical hierarchical tree structure, and its major team/personnel structure. This helps with resource allocations, based on skills required for each major organizational breakdown.

The OBS is normally broken down several levels and matched with the WBS to relate development teams/personnel to major development needs. Figure 8.5 illustrates an OBS based on Fig. 8.3, the House Cleaning Service WBS. Table 8.1 illustrates the mapping of the WBS to OBS.

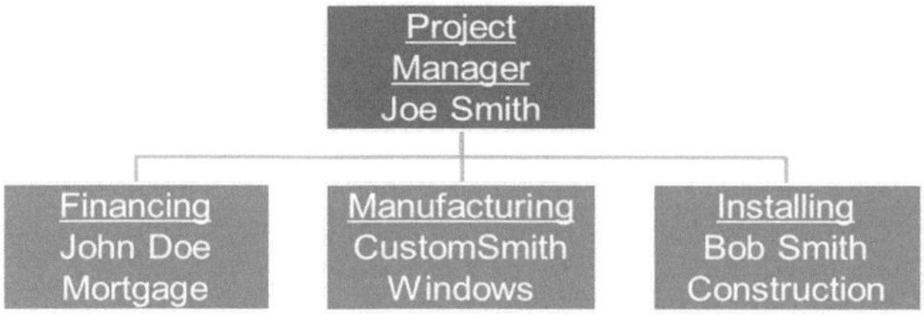

Fig. 8.4 High-level organization breakdown structure example

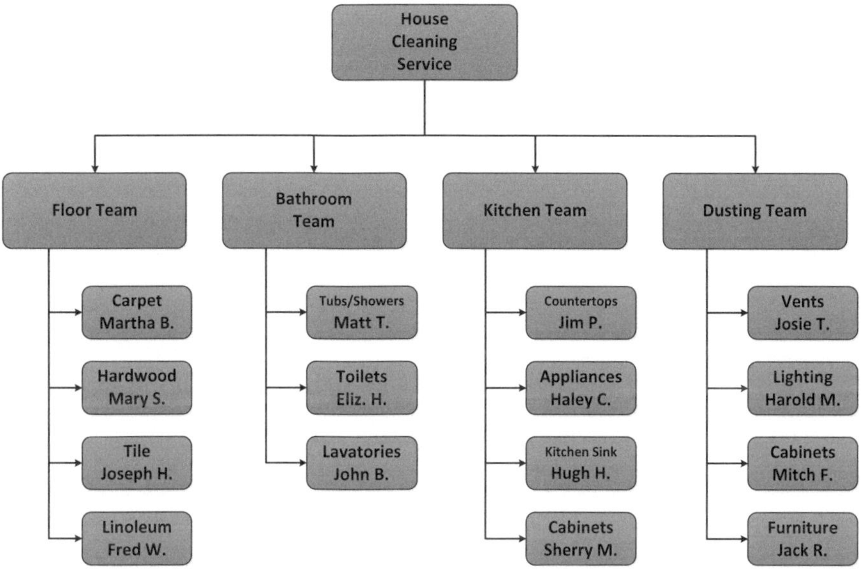

Fig. 8.5 OBS based on the WBS in Fig. 8.3

8.2 High-Level Program Plans

At a high level, System of Systems Engineering involves development of various program plans that will be used to direct and manage the design, implantation, and test the overall development program. What follows is an illustration of the Multidisciplinary Systems Engineering process involved in development of each of the top-level system engineering plans (e.g., System Engineering Management Plan—SEMP). System Engineering is an iterative, feedback-driven process that involved many disciplines that must cooperate and interact in order to ensure program success. The purpose of these process flows is to illustrate the interdependencies between each of the plans and disciplines that must take place over the course of time throughout the project.

Table 8.1 OBS—WBS Mapping for House Cleaning Service

	OBS																		
WBS	Floor team	Bathroom team	Kitchen team	Dusting team	Elizabeth	Fred	Haley	Harold	Hugh	Jack	Jim	John	Joseph	Josie	Martha	Mary	Matt	Mitch	Sherry
Clean floor		X																	
Carpet																X			
Hardwood																	X		
Tile														X					
Linoleum							X												
Clean bathrooms			X																
Tubs and showers																		X	
Toilets						X													
Lavatories													X						
Clean kitchen				X															
Countertops												X							
Appliances								X											
Kitchen sink										X									
Kitchen cabinets																			X
Dust house					X														
Dust furniture											X								
Dust vents															X				
Dust lighting									X										
Dust cabinets																			X

8.2.1 *Systems Engineering Management Plan (SEMP)*

In the development of complex systems, it is essential that all key participants of the system development process understand not only their own responsibilities, but also how they interface with each other. This is usually accomplished through the preparation and dissemination of the System Engineering Management Plan (SEMP).

The most important function of the SEMP is to ensure that all of the systems engineering personnel on the program (subsystem managers, design engineers, test engineers, systems analysts, etc.) know their responsibilities to each other. The SEMP, even though base-lined, is a living document, which contains a detailed statement of how the systems engineering functions are to be carried out in the course of system development for the program. Figure 8.6 illustrates the SEMP development process.

The SEMP development process flows as described next.

8.2.1.1 A0: Customer Inputs for SEMP

These inputs are shown as customer inputs because often these design artifacts are provided by the customer to guide/dictate the overall goals and objectives of the system design. The customer can be upper management, another company, or the Department of Defense as an example. If none of these provide these inputs, the Multidisciplinary Systems Engineering team should develop them, and make sure they are vetted with the customer, management, stakeholders, or users, whichever is deemed appropriate given the scope and need of the overall system. These inputs are:

- **Statement of Work (SOW)**: the SOW defines the overall scope of the development effort. The SOW captures the activities, system deliverables, and a basic time frame in which the system must be delivered. The SOW may provide detailed requirements, but there may be a separate requirements document.
- **Systems Engineering Plan (SEP)**: the SEP is an ever evolving document that captures the strategy for Multidisciplinary Systems Engineering (MDSE) across the System of Systems development. The SEP defines the relationship between MDSE and the rest of the Program. The SEP guides the technical aspects of the system architecture and design.
- **Test Engineering Management Plan (TEMP)**: the TEMP guides and directs that testing activities and events across the system development, formal integration testing, and formal factory sell-off testing, and transition testing for the overall System of Systems program. It is designed to describe the overall objectives and structure for all levels of Test and Evaluation activities.

8.2.1.2 A1: Program Management Plan (PMP) Development Inputs

As the name implies, the PMP describes the details of the planning of the System of Systems engineering project. It lays out the basic organizational structure of the program and provides the management processes and methodologies that will be

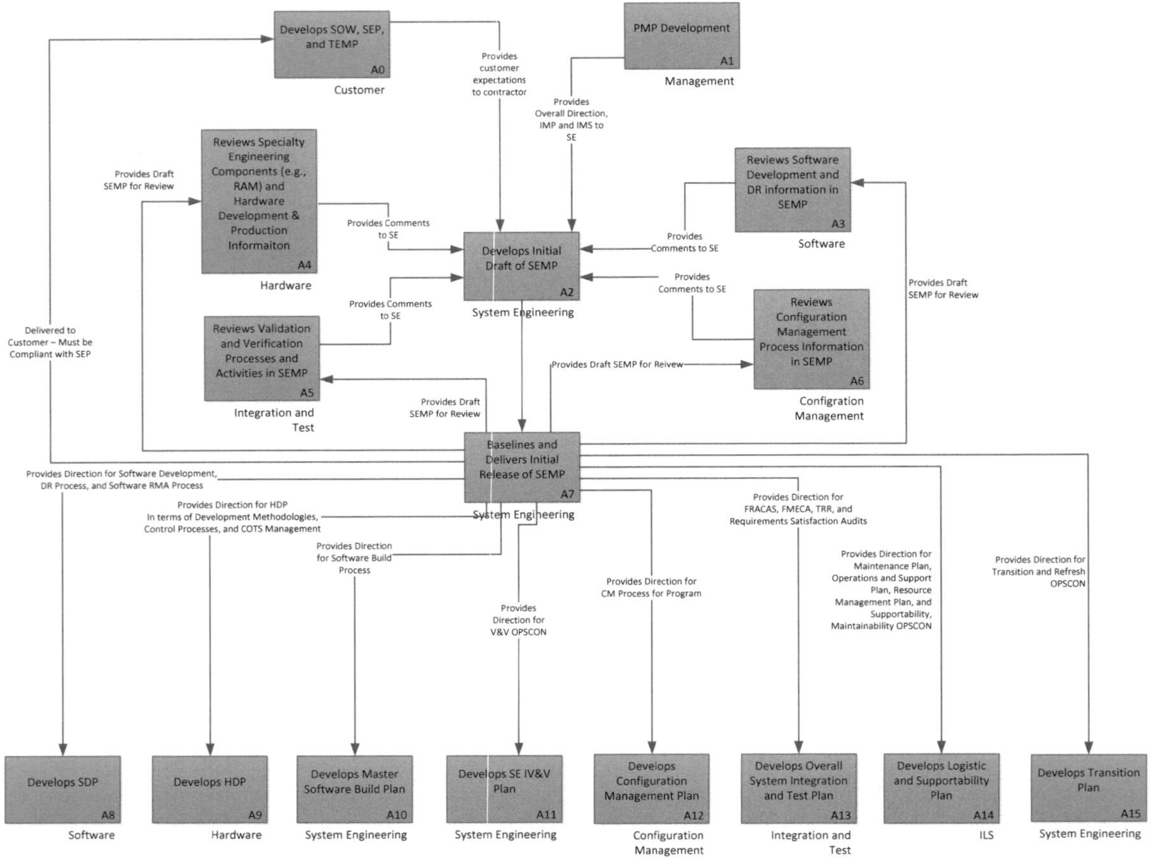

Fig. 8.6 Development of the Systems Engineering Management Plan

utilized throughout the system development efforts. Like the SEMP, the PMP is a living document that should grow as the program grows and changes. The PMP drives program execution, monitoring, process control, and how the program will be closed out when completed.

8.2.1.3 A3: Software/Hardware Development and Deficiency Report (DR) Structure Inputs

These inputs into the SEMP describe how software will be controlled, how changes to software and hardware are structured and maintained; traceability of the developed software/hardware and how software/hardware deficiencies ('bugs') are handled throughout the development efforts. This includes methods for version control as well as change control management.

8.2.1.4 A4: Specialty Engineering Inputs

Specialty Engineering includes engineering disciplines that are a major part of Multidisciplinary Systems Engineering, but are not thought of as part of main stream engineering. This includes disciplines like System Safety, System Security, Human-System Interface (HSI), Quality Assurance (QA), Reliability, Maintainability, and Availability (RMA), and others. While these disciplines are not considered to be the major system considerations (unless the SoS is a safety system), integration of these disciplines into the overall System of Systems architecture and design is essential and should be a major objective of Multidisciplinary Systems Engineering. Many of the Specialty Engineering system requirements become constraints on the overall system design, e.g., mean-time-to-repair. One of the major roles of Multidisciplinary Systems Engineering is to balance these constraints against the overall systems architecture to ensure all requirements are met and the overall System of Systems provides the system performance required while maintaining the overall utility and usability the system is required to attain.

> ***Question to think about***:
> *What happens to the development, when specialty engineering is not taken seriously by the development team?*

8.2.1.5 A5: Verification and Validation Structure Inputs

The Verification and Validation methodology for the System of Systems program is important to understanding how the system will be tested. Multidisciplinary Systems Engineer should take a page from both Vladmir Lenin's and President Ronald Reagan's book:

Doveryai, no Proveryai (Trust but Verify)

Verification and Validation (V&V) is an important discipline for the Multidisciplinary Systems Engineer, as this determines if the system design complies with the requirements, objectives, constraints, and specifications (verification), while validating that the as-built system is viable (constitutes and operational mission/business system). You might think of it this way:

- Verification: Are we designing and implementing the system correctly?
- Validation: Are we designing and implementing the correct system?

Verification and Validation is crucial to the overall success of a System of Systems program as these functions serve to find errors that result from hardware and software faults. The question the Multidisciplinary Systems Engineer must ascertain is how much V&V is requirement for a given system design. The level of V&V drives the overall testing plans for the system, throughout the development and system sell-off life cycle.

8.2.1.6 A6: Configuration Management Plan Inputs

Configuration Management is the formal capture and recording of information that describes the System of Systems architecture, hardware and software designs, test plans, transition plans; all of the formal design and test products across the system development life cycle. This includes version control of all documentation, all hardware and software updates (e.g., which version of Java is being used), location and IP addresses of hardware devices, ports, protocols, interfaces (internal and external); all information required for operations and maintenance of the system, once delivered and turned over to the customer. This is required for the Multidisciplinary Systems Engineer to make informed decisions about proposed changes to the system, to ensure the proposed changes will not adversely affect other components, subsystems, services, or elements of the overall System of Systems.

8.2.1.7 A2: Develop Initial Draft of the Systems Engineering Management Plan (SEMP)

Utilizing all of the inputs described above, a draft of the SEMP is written and sent out to all development groups with the program for review and feedback before the final SEMP is established and put under Configuration Management (CM) control. As discussed, the SEMP describes the Multidisciplinary Systems Engineering's plans, controlling structures, and how they program intends of conducting and delivering a fully integrated systems. This includes:

- Describing technical planning and technical management of the program.
- Describing how the technical plan and the project plan are in sync.
- Defining the technical activities required to complete the development program.
- Describes the flow of Multidisciplinary Systems Engineering activities required across the program life cycle.

- Describes the technical views, measures-of-effectiveness (MOEs), measurements, and performance metrics required to assess the program.
- Describes how the SEMP will be utilized to control the scope of the overall execution of the System of Systems program.

8.2.1.8 A7: Baseline and Deliver Initial Release of the SEMP

Once the draft version of the SEMP has been adequately reviewed, and all issues have been adjudicated, the SEMP is base lined and put under Configuration Management control. Hoever, you must understand that the SEMP is a living document and will be updated and changed as the System of Systems development efforts moves through the implementation, test, and delivery. Once a SEMP is base lined, the SEMP information is utilized as inputs to a number of other program plans that are developed. Sections 8.2.1.8–8.2.1.15 describe these plans.

8.2.1.9 A8: Develop the Software Development Plan (SDP)

The Software Development Plan (SDP) defines the overall software development activities and phases required for implementing the software required for the System of Systems. The SDP is important whether the software development methodology is Waterfall, Iterative, Agile, or eXtreme; just because you are agile, doesn't mean you don't have a plan! The SDP includes, but is not limited to:

- Overview of the scope, objectives, and purposes for the System of Systems software.
- Software Deliverables.
- Organizational Structure for the software teams.
- Major phases and milestones for the software development, integration and test activities, including how progress and software quality will be measured [108].
- Overview of the software development methodologies (e.g., iterative, agile, etc.), including the tools and techniques to be utilized throughout the software development and test life cycle.
- Configuration Management of the software.

8.2.1.10 A9: Develop the Hardware Development Plan (HDP)

Similar to the SDP, the Hardware Development Plan (HDP) lays out the requirements, standards, methodologies, and development activities to develop the hardware environment(s) for the System of Systems design. This includes definition of a common terminology, acronyms, and vocabulary for the hardware development efforts. Since often, hardware designs includes Commercial Off-the-Shelf (COTS) hardware, the SDP includes expectations for hardware acquisition as well as expectations about the H/W development processes and which H/W design drawings will be utilized across the program. Figure 8.7 below depicts (at a high level) the Hardware Development Process.

8.2.1.11 A10: Develop the Master Software Build Plan

The Master Software Build Plan provides a phased approach to how the overall software structure of the program will be handled. This includes a build schedule for when code will be written, tested, and delivered for which capabilities, subsystems, services, etc. across the software development life cycle for the System of Systems program. The Multidisciplinary Systems Engineer must be cognizant of these plans and understand the ramifications in order to plan and manage the overall System of Systems development across the program. This Master Software Build Plan becomes a major input to the overall System of Systems Integrated Master Schedule and Integrated Master Plan.

8.2.1.12 A11: Develop the System Engineering Verification and Validation Plan

Based on the initial Verification and Validation structure and the information in the SEMP, the Verification and Validation Plan (V&V Plan) is developed. The V&V Plan describes the processes, metrics, and how software deficiencies (i.e., deficiency reports) are handled throughout the software development life cycle [87]. V&V is the processes of quality control on the software development, ensuring that the developed software meets all specifications, requirements, constraints (e.g., reliability), and that the software meets the system's intended purpose. This V&V Plan includes the methodologies that will be utilized for testing each of the software requirements [109].

8.2.1.13 A12: Develop the Configuration Management Plan (CMP)

Based on the original Configuration Management inputs and inputs from the SEMP, the Configuration Management Plan (CMP) is developed. The CMP established the integrity of hardware, software, systems, and test development and execution across all elements of the System of Systems development and transition efforts. Configuration Management (CM) is an integral part of the overall disciplines the Multidisciplinary Systems Engineer should be familiar with. The CMP includes, but is not limited to:

- How program artifacts are named
- The CM process across the program
- The change control process for artifacts under CM control
- How versions of artifacts are tracked
- How CM tools (usually COTS) are utilized to provide CM functionality

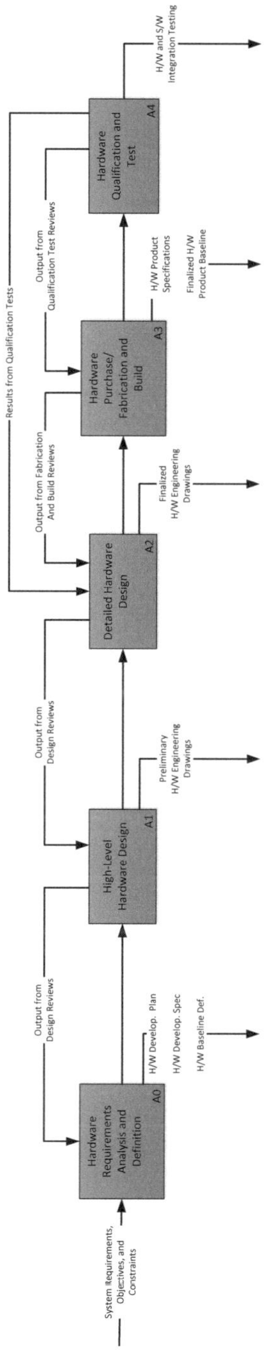

Fig. 8.7 Hardware development phases

8.2.1.14 A13: Develop the System Integration and Test Plan

Information on Integration and Test from the SEMP is utilized to create the overall program Integration and Test Plan (ITP). For the Multidisciplinary Systems Engineer, integration and test planning must be an integral part of the overall System of Systems design. The ITP provides the definition of integration testing that focuses on system functionality; how the elements of the System of Systems integrate together to form an overall functional, viable, usable system. This includes the identification of functional "threads" that test a particular use of the system throughout each element, subsystem, service, and component of the system that is utilized for that system functionality.

8.2.1.15 A14: Develop the Logistic and Supportability Plan

The Logistic Support and Supportability Plan should be an extension of the SEMP that describes the overall Multidisciplinary Systems Engineering strategies for functional support and to guide the System of Systems development life cycle through quantifying life cycle costs; looking for ways to lower costs and decrease the overall program logistics footprint, which in turn makes the System of Systems easier to support and maintain. Logistics support is truly a Multidisciplinary Systems Engineering discipline, requiring activities including, but not limited to:

- Reliability, Maintainability, and Availability (RMA)
- Supply Chain Engineering (description of spare parts)
- Training (maintenance and operations training)
- Technical Writing (operations and maintenance manuals)
- Facilities (physical footprint, power, HVAC, etc.)
- Packaging and Handling (how to pack up, transport, deliver, and set up the system)
- Computer Support (systems administrators, database administrators, etc.)

8.2.1.16 A15: Develop the System Transition Plan

The Transition Plan facilitates the transition of the System of Systems from a development program to an Operations and Maintenance program. System Transition should be (but most often is not) an integral part of the overall design. For the Multidisciplinary Systems Engineer, how the system will be taken from development to operations must be designed into the overall architecture and development. This includes the controls, reporting procedures, as well as the overall operational risks of the System of Systems. In addition, there are optional disciplines that might be included, depending on the scope and objectives of the SoS. This includes plans like the Operational Network Plan if the SoS is a network intensive or globally distributed system.

8.3 Systems Engineering Logistics and Support Concept Development

In the following sections we expand upon those plans and designs that are of major importance to developing, testing, and deploying a SoS. The Logistic Support facilitates development and integration of the logistics support elements listed below. It provides the concepts necessary to specify the design, development, acquisition, test, fielding, and support of the system. The Logistics and Support elements are:

1. Maintenance Planning.
2. Supply Support.
3. Support and Test Equipment/Equipment Support.
4. Manpower and Personnel.
5. Training and Training Support.
6. Technical Data.
7. Computer Resources Support.
8. Facilities.
9. Packaging, Handling, Storage, and Transportation (PHS&T).
10. Design Interface.

All elements of the system must be developed in coordination with the overall systems engineering effort. Tradeoffs may be required between elements in order to acquire a system that is affordable (lowest life cycle cost), operable, supportable, sustainable, transportable, and environmentally sound within the resources available. Figure 8.8 illustrates the System Logistics and Supportability Concepts Development Process.

Question to think about:
When should the logistics and supportability plan start to be exercised during development?

8.4 Software Engineering: The Master Software Build Plan

For each Software Requirement Specification (SRS) requirement, the Master Software Build Plan establishes when in the software development schedule that requirement is met (or when the software is completed) [110]. This is not in terms of calendar dates, but in terms of internal program milestones (iteration builds, etc.). Figure 8.9 illustrates the Master Software Build Plan development process.

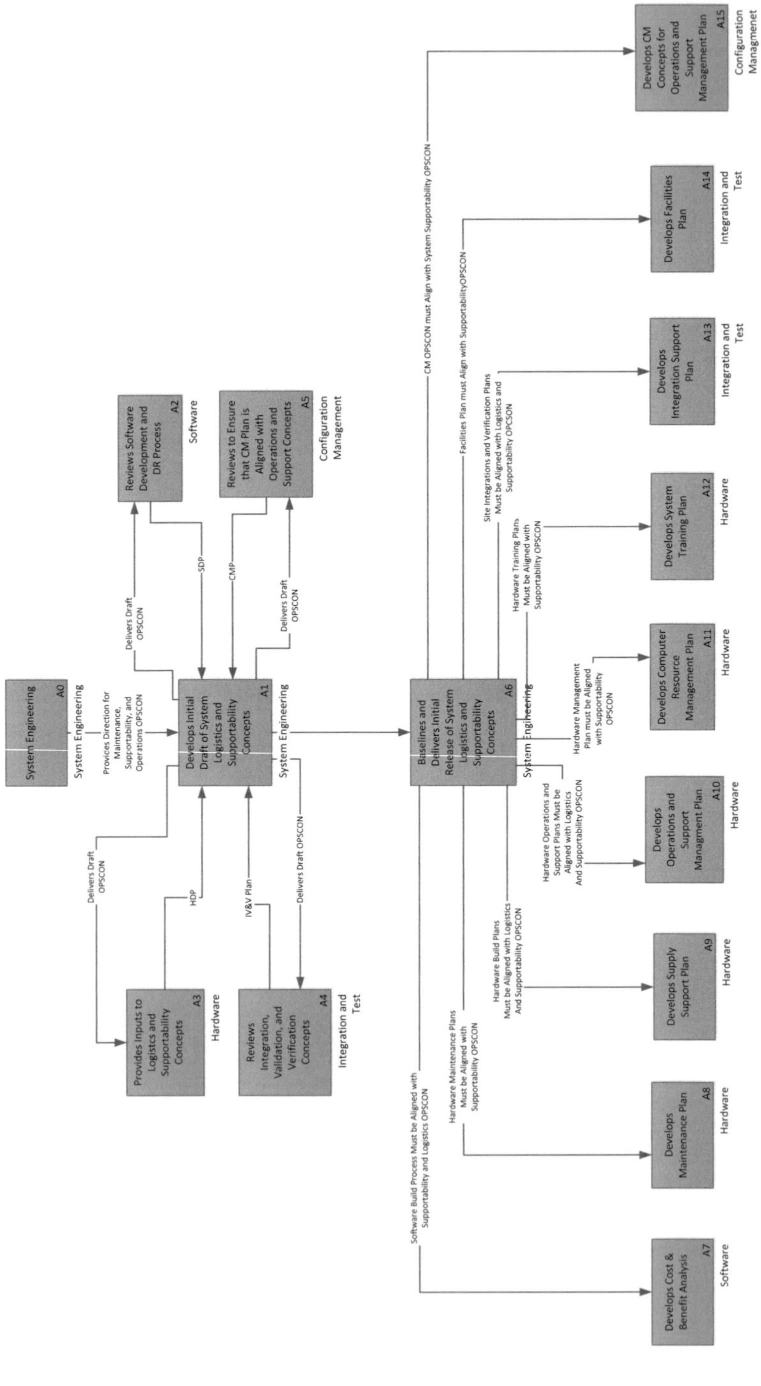

Fig. 8.8 System logistics and supportability plan

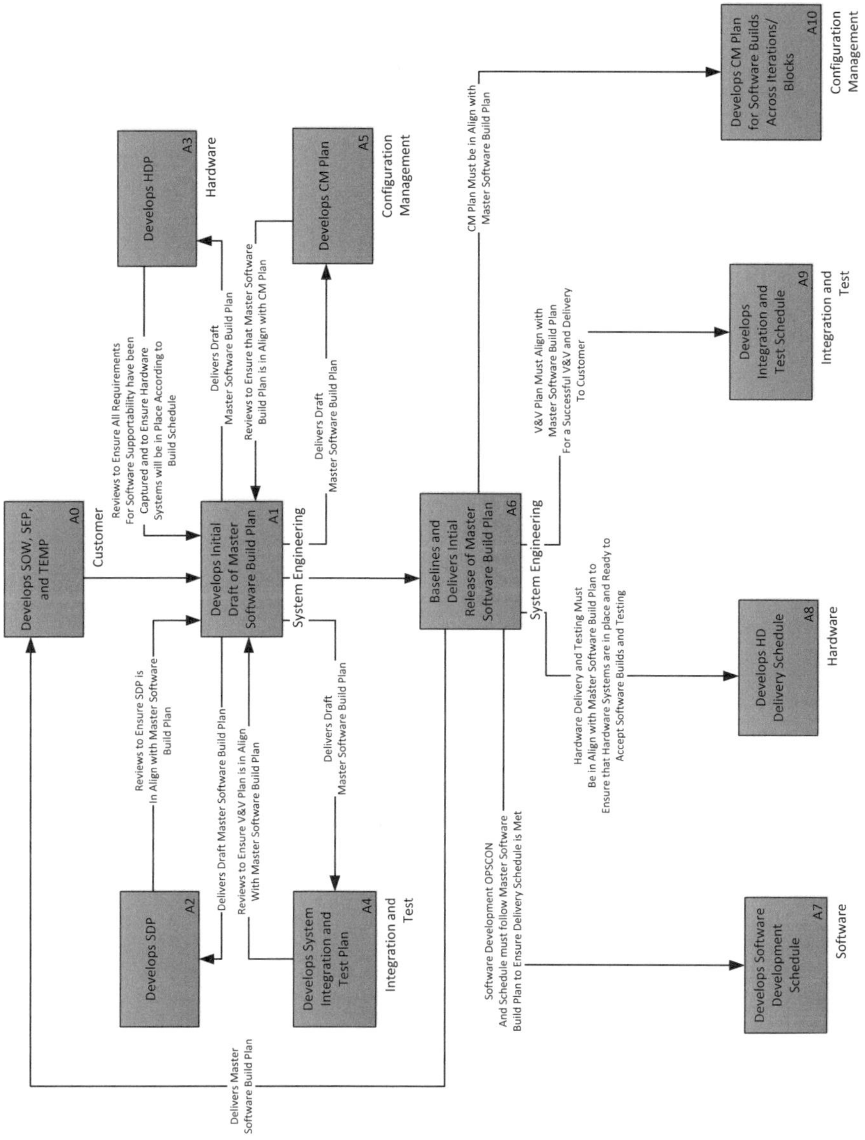

Fig. 8.9 Software master build plan development

8.5 Transition Plan Development

The Transition Plan is used to describe how deliverables of the project will be brought to full operational status, integrated into ongoing operations and maintained. Figure 8.10 illustrates the Transition Plan development process.

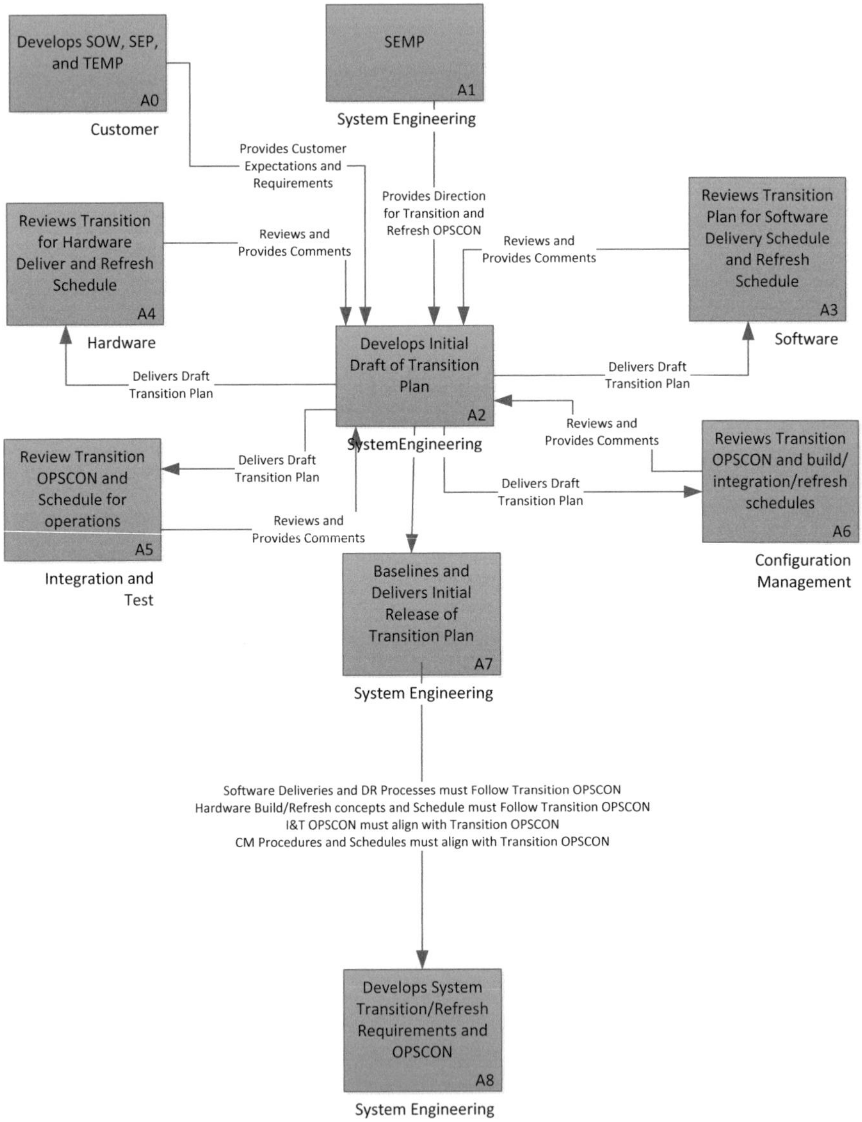

Fig. 8.10 System transition development plan

8.6 Information Assurance Plan Development

The Information Assurance Plan defines the technologies, processes, and procedures the program will use to protect the confidentiality, integrity and availability of information on the program, based on various government specified security regulation that must be followed. Figure 8.11 below illustrates the Information Assurance Plan development process [111].

8.7 System Safety Plan Development

The primary purpose of the System Safety Plan is to establish the organization and define activities to identify possible hazards (risk of damage, injury, or death) and to analyze and reduce the risk of their occurrence. This plan includes plans for safety of the system, that is, avoidance of danger to life, health or property while in use including storage, transportation, and test. Initially, a 'Preliminary Hazard Analysis' is performed to determine the potential hazards that may occur as a result of the system. Figure 8.12 below illustrates the System Safety Plan development process [112].

Questions to think about:
What is the relationship between the System Safety Plan, the Software Development Plan, and the Hardware Development Plan?
How does the System Safety Plan affect Design for Maintenance and the Logistics Plan?
Hint: *it is important*!

8.8 Operational Site Activation Plan Development

The Operational Site Activation Plan includes all Facilities, materials, plans, procedures, and training required to install and activate (bring into operations) the system at the customer site. Figure below illustrates the development process for the Operational Site Activation Plan. Figure 8.13 illustrates the process of developing a Site Activation Plan (SAP).

Question to think about:
Should the Site Activation Plan address Data Management?

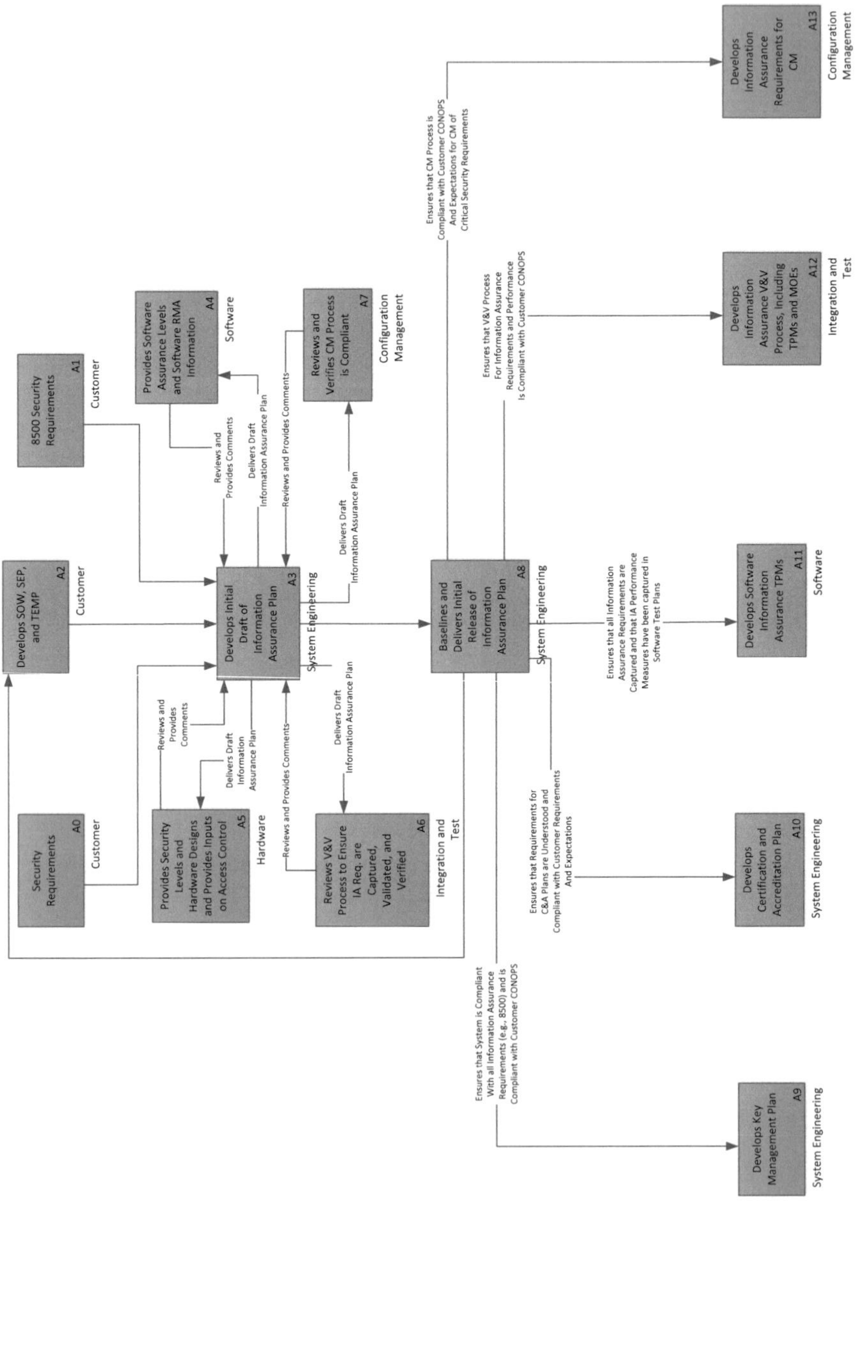

Fig. 8.11 Information Assurance Plan development

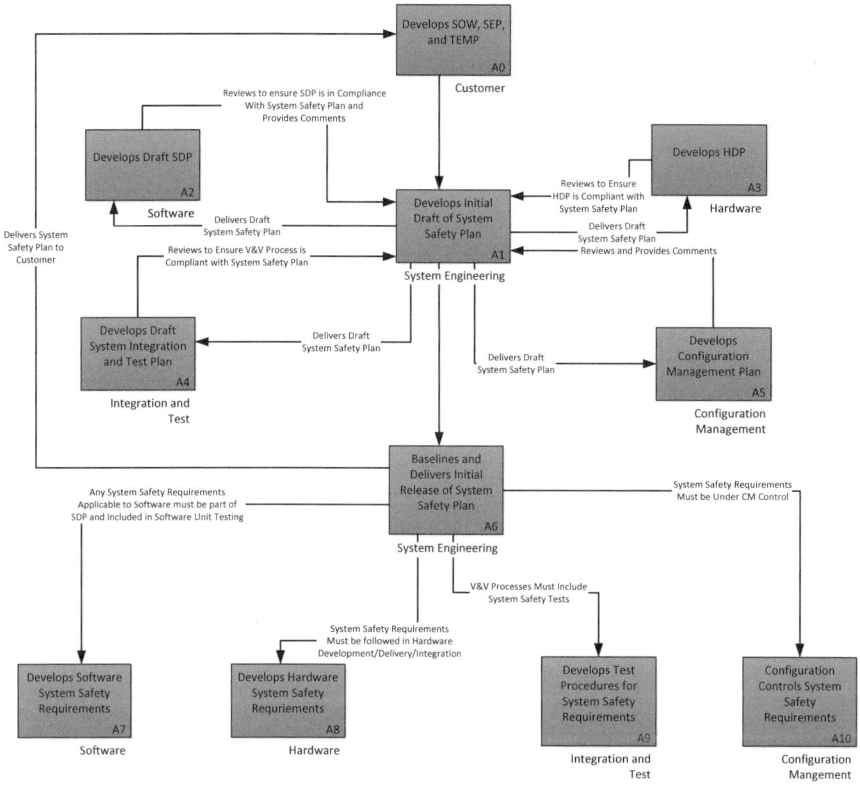

Fig. 8.12 System Safety Plan development

8.9 Facilities Development Plan

Facilities are permanent or semi-permanent assets required to support the system that includes conducting studies to define types of facilities required (or facility improvements), locations, space needs, environmental requirements, and equipment. Facilities may include:

- Modifications or rehabilitation of existing facilities used to accomplish training objectives.
- Facility Design.
- Interconnecting Cabling.
- Utilities.
- Environmental Control.
- Construction.
- Facilities Acceptance.
- Supervision, inspection and overhead costs to design, construct/arrange, and accept the facility.

Figure 8.14 illustrates the development of the Facilities Plan.

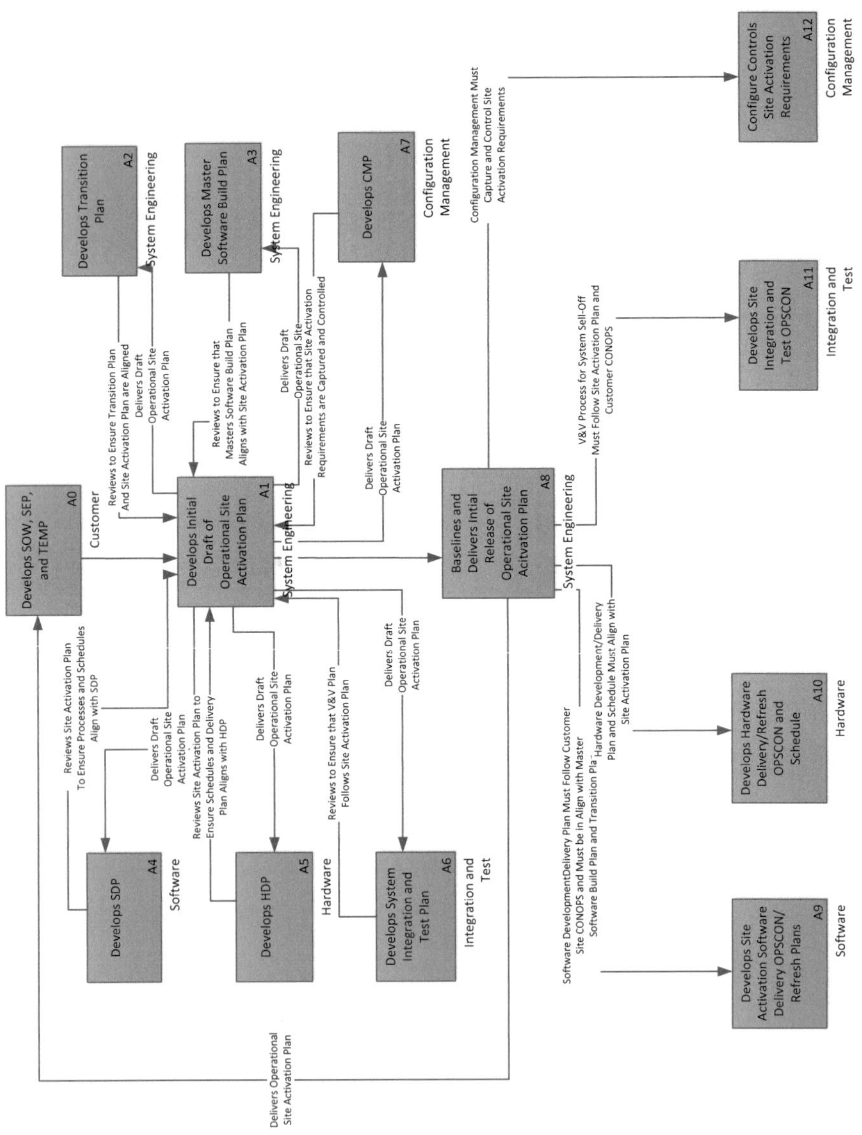

Fig. 8.13 Site Activation Plan development

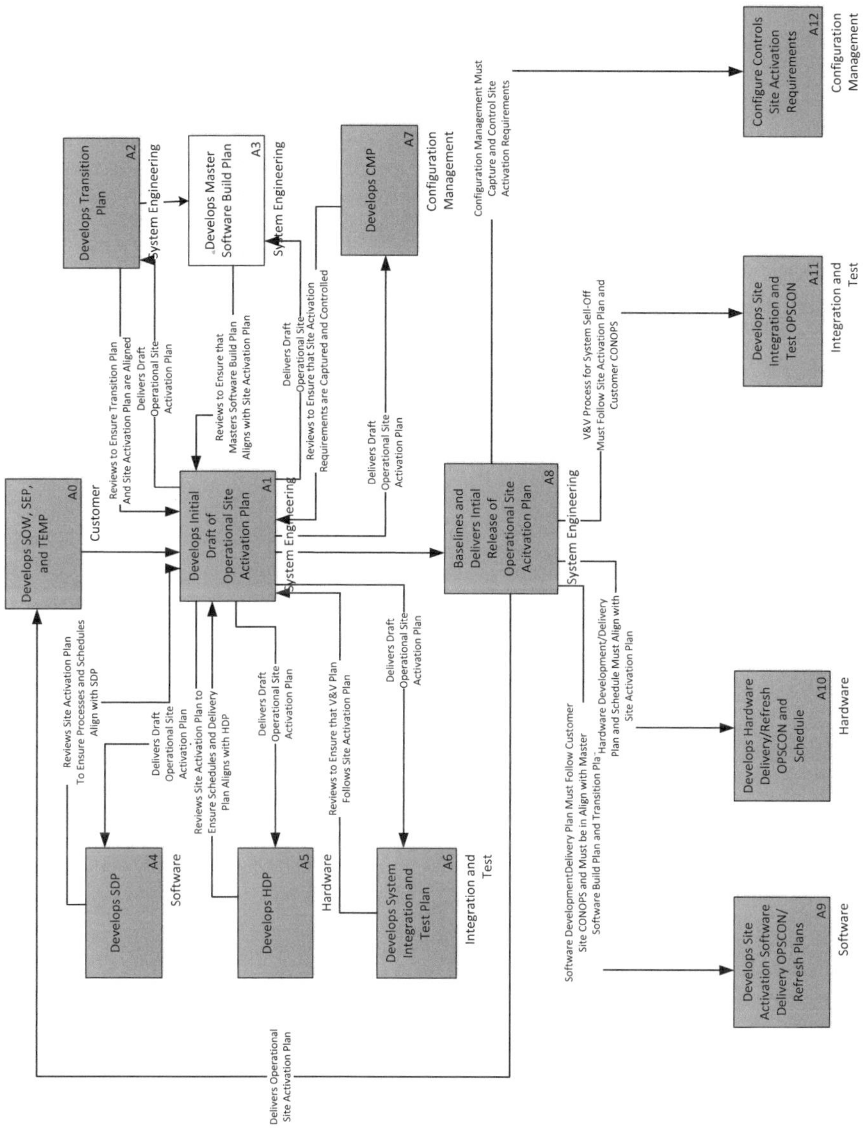

Fig. 8.14 Facilities Plan development

8.10 Systems Engineering IV&V Plan

The System Engineering Integrated Verification and Validation Plan describes the system engineering efforts required to provide integrated requirements validation and verification that the system meets all validated system requirements. This integrated test plan must cover the entire domain of system test, encompassing everything from design verification through field maintenance of the system. The test plans integrate with all system, subsystem, and component level process. The test plans include all embedded and off-line testing procedures. The IV&V Plan provides an integrated total test solution for the system, defining which verification methodology will be used for each system, subsystem, and component level requirement. Figure 8.15 illustrates the IV&V Plan development.

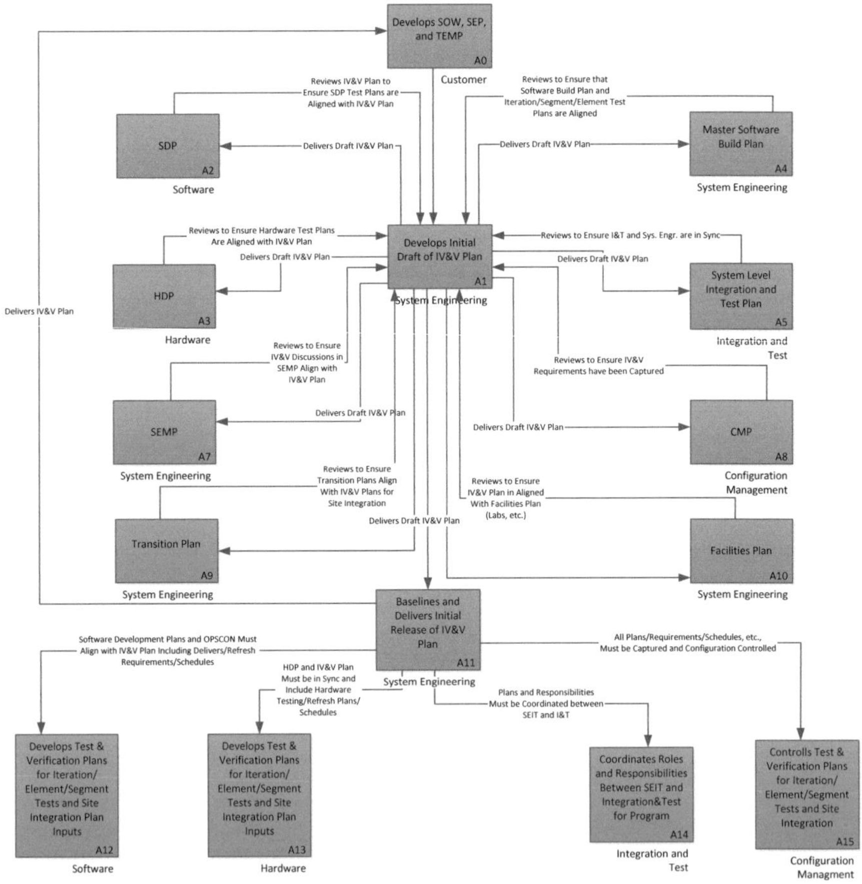

Fig. 8.15 Integrated Verification and Validation Plan development

8.11 Human Engineering Plan Development

The Human Engineering Plan describes the program efforts to ensure compliance with Human Factors requirements provided by the customer and all applicable Human Factors Design Guides with which the program may be required to comply. This plan describes the human engineering trade studies, tests, mock-up evaluations, dynamic simulations, and design/system reviews that will be required to ensure design and performance compliance across the program. Figure below illustrates the Human Engineering Plan development process. Figure 8.16 illustrates the Human Engineering Plan development.

8.12 Case Study Multidisciplinary Systems Engineering and System of Systems Complexity Context

We have spent much time discussing Multidisciplinary Systems Engineering (MDSE) and how it applies to a System of Systems (SoS) architecture design and implementation. MDSE should touch and drive every aspect of the SoS design and be part of

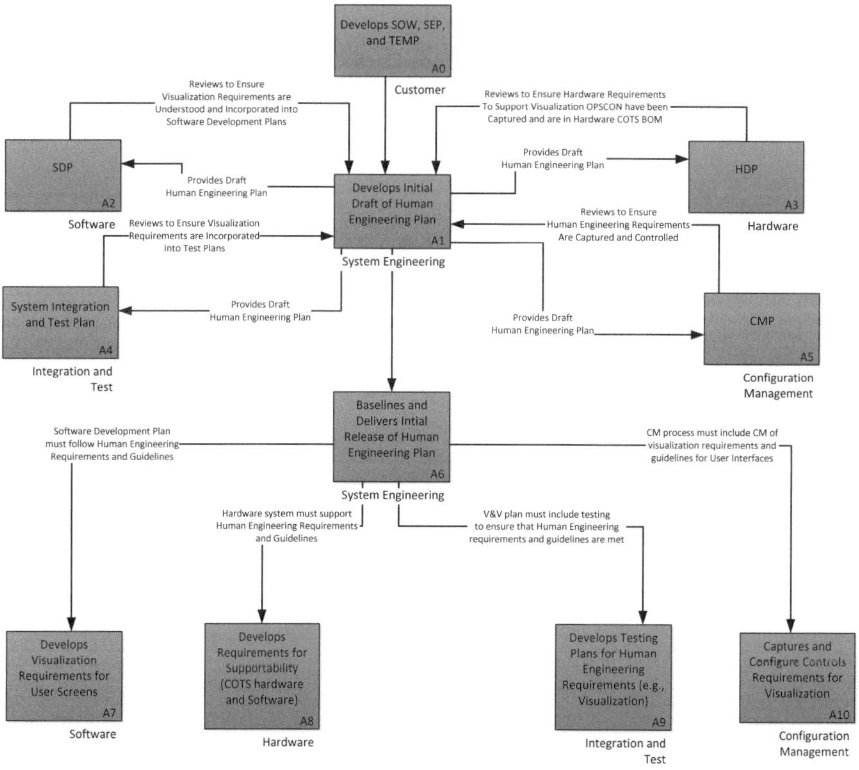

Fig. 8.16 Human Engineering Plan development

each discipline involved in the design, implementation, testing, integration, V&V, deployment, and transition. Figure 8.17 provides the overall context diagram for MDSE as it applies to the overall life cycle of a SoS. Refer to the acronym list provided at the end of the book for any acronyms the reader is unfamiliar with [113].

As depicted in Fig. 8.17, MDSE encompasses a host of engineering and management disciplines. Each of the plans shown in Fig. 8.17 require knowledge and input from the MDSE personnel. This drives home the fact that the human element is most important. The human role in System of Systems engineering makes or breaks a system design, and one must understand human behavior, human decision making, as well as overall SoS technological complexity, in order to better understand the sources of complexity within MDSE and in the design of SoS architectures. This understanding of human behavior and its role in MDSE will, we believe, enable better overall systems designs and success in the design, implementation, integration, test and deployment of SoS.

8.12.1 Human Behavior and System of Systems Design

System complexity theory [59] tells us that complexity characterizes systems with multiple components which interact with each other in possibly linear and non-linear ways. Factors of complexity include, but are not limited to:

- **Physical System Complexity**: this involves measuring the probability of a given system state vector. This mathematical measure dictates that any two distinct system states are never considered the same and represent different possible state of the system.
- **Information Complexity**: this measures the number of possible information properties across the system that can be transmitted and observed. A complete collection of these properties is called a system state.
- **Network Complexity**: the number of connections and connection combinations between elements/component of the system.
- **Software Complexity**: the number of possible interactions between the software components of the system. This measures the complexity of the software design and is distinct from computational complexity.
- **Perceptual Complexity**: this is a measure of human operator complexity equal to the square of the number of visual or operator accessible features, divided by the number of system components which the human operator may interact with. Such complexity affects human behavior and human decision making.

The overall complexity of a system depends on the interaction of all of the complexity components described above and the emergence and self-organization driven by these complexity components. When the overall SoS architecture design contains a heterogeneous collection of system elements and components, the complex interactions between these elements will drive complex behaviors peculiar to human behaviors. A study conducted by the Air Force [114] identified the roles of

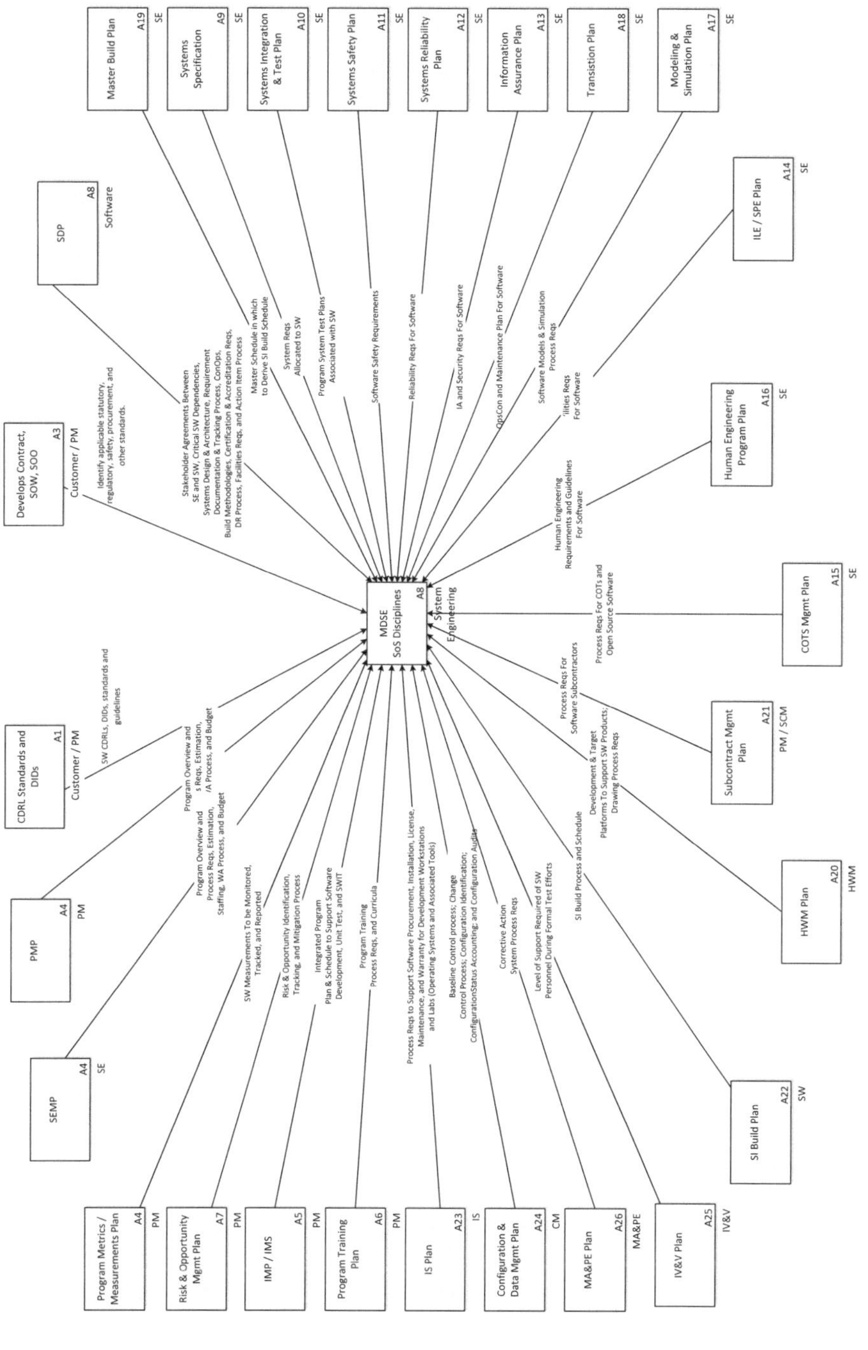

Fig. 8.17 Multidisciplinary Systems Engineering discipline context diagram for system of systems

human operators and their perceived complexity as crucial to successful operations and maintenance of delivered SoS. The study stated [114]:

> Whenever the Air Force generates a System of Systems, interaction among the systems often includes human-to-human interactions. If the machine-to-machine aspect of the SoS is weak, then it falls on humans to achieve the interaction. This can, and often does, create a very challenging environment for the humans; sometimes leading to missed opportunities and serious mistakes…

Human behavior and decision making through processes must be taken into account as sources of complexity within a SoS context. The MDSE must evaluate the human-human and human-machine interactions in order to reduce the human-system interoperations in order to ensure a User-viable, operational SoS. Effective MDSE designs should carefully consider Human Systems Engineering in the overall SoS architectural design and SoS implementation. Capturing the potential human-system interactions at various levels within the SoS design are essential to understanding the system reliability, risk assessment, and maintainability [115]. What this means for MDSE is that there must be collaboration between the MDSE architects and designers with human operators, along with Human Systems Engineering professionals in order to design a SoS that not only meets all requirements and objectives, but is useable by operators that have a level of expertise commensurate with customer expectations. Designing a SoS that requires PhDs to operate, when the customer is expecting a SoS that can be operated by high-school graduates will not be accepted.

8.13 Discussion

Of the entire MDSE life cycle process, creating program plans may seem like the least interesting of all the tasks associated with a System of Systems life cycle. However, without them, the odds on delivering the right product are slim. Realize that only those plans that are required for a given System of System should be produced. Also realize, as discussed, that "the plan" may be a paragraph or two in a combined set of plans, or may be a lengthy document. Many will push for process for process sake, assuming process will save the program. The MDSE must resist this, as sound multidisciplinary systems engineering is what is required, not process. Process is necessary, but not sufficient, to bring a viable system to fruition.

Chapter 9
Plan Development Timelines

In Chap. 8 we described the processes, inputs, outputs, and dependencies of various disciplines and development plans. However, another aspect of MDSE plan development is the information on the phasing or timing of when each plan is required across the system development life cycle. This chapter lays out each major program milestone and illustrates which plans must be in "initial release" maturity and "baseline" maturity during each major phase of program development. The arrows indicate plan dependencies and phasing across program milestones. The point to be made is that plans are not made in a vacuum and require cooperation among all MDSE and management disciplines to ensure that the plans are complete and meet the program needs across the entire program lifecycle. One major point to understand about the plans discussed in Chap. 8 and the timelines presented in Chap. 9, is that the amount of detail needed in each greatly depends on the project size and scope. In some cases these plans are separate documents unto themselves. In other cases they may be a paragraph in an agile Sprint plan that discusses changes as a result of the current plans. But, for the MDSE, these issues must be thought through throughout the SoS design, development, integration and test, and deployment phases. Figures 9.1 and 9.2 illustrate this. Figure 9.1 illustrates a traditional program development cycle [21]. For the development process depicted in Fig. 9.1, the engineering plans are finalized early in the development process and only changed when major program events dictate the changes.

For the process in Fig. 9.1, requirements are allocated and/or derived, program plans are made, budgets and schedules established, and you march forward [116]. For this development cycle, change is bad, for it causes rework of the entire development cycle, driving up cost and schedule. And, the farther into the development cycle you are when you discover problems, or changes happen (new or changing requirements), the costlier they are to "fix" both from a cost and schedule perspective [116].

In contrast to Fig. 9.1, Fig. 9.2 depicts the Agile Program Development Life Cycle. In Fig. 9.2 we see the typical Sprint rhythm of initiating, planning, and execution of agile Sprints during the agile development process.

J.A. Crowder et al., *Multidisciplinary Systems Engineering*,
DOI 10.1007/978-3-319-22398-8_9

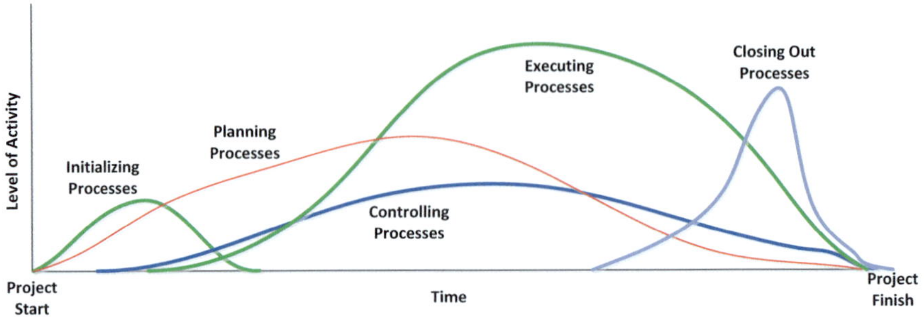

Fig. 9.1 Traditional program development cycle

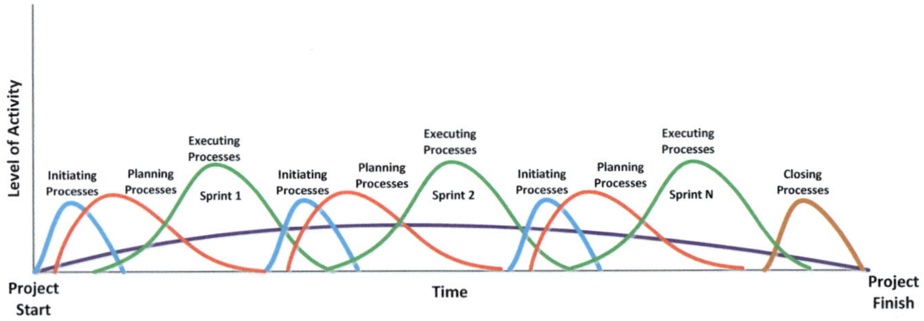

Fig. 9.2 Agile system development process

The length of the Sprints is dependent on the program/project, the complexity of the software, total duration, and team membership. The duration of the Sprints is not relevant to this discussion. One of the big advantages to agile development is due to the adaptive nature of the Sprints. If a problem is discovered it can be rolled into the next and subsequent Sprint planning sessions and rolled into the schedules and does not require a major re-plan of the entire project, the way it does in classical development. For this development process, project planning (e.g., System Architecture, Software Architecture, design, integration and test, etc.) are subject to change as the program progresses [117]. Normally, because there is working software after each Sprint, problems and required changes are discovered earlier in the development process than in conventional or classical development methodologies. Customer and management feedback after each Sprint provide the opportunities to re-vector the development efforts before major cost and schedule have been expended. This allows the program to adjust design documentation as the program progresses, such that the "build-to" documentation and the "as-built" system agree.

The remainder of Chap. 9 is devoted to laying out the process flows, in regards to program and technical plans. Again, these may be full documents (e.g., SEMP), or they may be paragraphs depicting the evolution of these plans (e.g., V&V) across the SoS development program as the agile design changes.

9.1 Authorization to Proceed (ATP)-to-Systems Readiness Review (SRR) Development

The first major milestone for any System of Systems development is the Authorization to Proceed (ATP) and the Systems Readiness Review (SRR).

9.1.1 Authorization to Proceed (ATP)

ATP or Work Authorization is a formal approval from the customer (whether internal or external) that activities required for the System of Systems development can begin. There is a major amount of pre-work required before ATP can be determined. Figure 9.3 illustrates the plan development and relative timing required for ATP and for SRR. Initial releases of the plans and documentation shown in Fig. 9.3 are considered drafts of these plans and must be scrutinized by the MDSE organization, along with the customer, to determine their correctness against the requirements, SOW, OPSCON, and other documentation (e.g., Statement of Objectives).

9.1.2 System Readiness Review (SRR)

The System Readiness Review (SRR) is performed to ensure the initial System of Systems architecture and designs are based on a mutual understanding (customer and MDSE team) of the overall system requirements and that the initial architecture concepts are reasonable and encompass the customer's needs. This review ensures that all planning (e.g., testing philosophy, logistics plan, hardware plans) are acceptable. The SRR provides a formal assessment to demonstrate that the documentation are consistent and technically viable. The plans produced in support of the SRR are important, as they drive cost and schedule estimation, based on their assessment of the requirements decomposition/derivation. The documentation illustrated in Fig. 9.1 provides the following:

- Initial requirements decomposition/derivation and allocation of requirements to hardware and software designs.
- Initial software service definition that adequately covers technical and non-technical requirements, as well as quality attributes and all deliverables.
- An initial System Engineering Plan (SEP), and System Engineering Management Plan (SEMP).
- An initial program Risk Assessment (both technical and non-technical risks).
- An initial assessment of Trade Studies that must be performed.
- An initial Verification and Validation plan.
- Initial definition of simulators and models required across the system development life cycle.

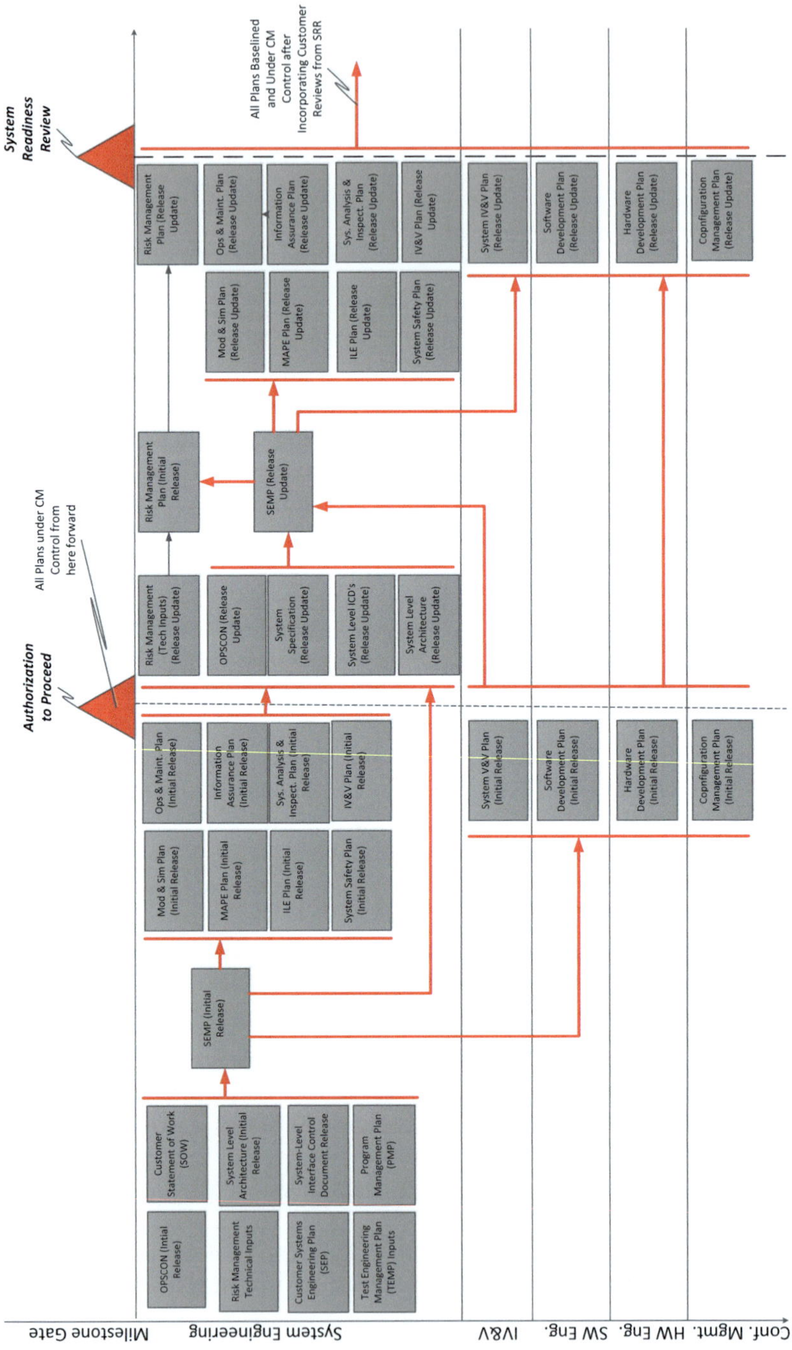

Fig. 9.3 Plan development through ATP and SRR

Questions to think about:
Is the SRR important?
If so, why are they held as early in the program lifecycle as they are?
Why are some of the critical plans being developed after the System Requirements?

9.2 System of Systems Element and Subsystem Design Development

As has been discussed throughout the book, SoS designs should take into account not only the context of the entire SoS, but should understand the context of the Element and Subsystem and how they will operate within the overall SoS Enterprise. This should include an incremental development strategy defined for each subsystem and element, as well as their incremental changes (or improvements) required for the SoS. SoS change over time in terms of operational environments, objectives, needs, and technological needs (i.e., how to infuse new technology into an existing SoS). Many SoS operate for decades where a technology and improvement plan must be built into the overall SoS architecture and designed to allow flexibility, adaptability, expandability, and modifiability throughout the full life cycle of the SoS. Resiliency and adaptability are becoming major components of all large-scale systems. This includes adaptability to changing external interfaces, which are driven by Interface Control Documents (ICDs), which specify interfaces between the SoS and external entities. These interfaces can be to the overall SoS, or can be an interface to a given element or subsystem. ICDs are used to describe the interface in terms of data/information, including the size, format, latency, use of the data/information and like factors. ICDs also define the interface protocols to be used, e.g., Secure File Transfer Protocol (SFTP). When ICDs are adequately specified, it allows development teams to test their subsystems/elements/SoS by simulating (or stubbing out) the interface. This applies to internal interfaces as well as external interfaces, although internal interfaces between SoS elements or subsystems are normally documented and handled through Interface Requirements Specifications (IRS). Modularity and interface abstraction leads to a more easily maintained SoS and provides extensibility, since what's on the other side of the interfaces can be upgraded or modified without affecting the other subsystem or element as long as the interface does not change. The concept of an ICD or IRS can exist without explicit documentation and (contrary to the beliefs of many) are still very useful for agile projects. The ICD or IRS need not be an explicit textual document, but may be as simple as a table of information givers and receivers which may evolve over time [36]. This can be implemented using a dynamic database to provide interaction information. Subsystems and Elements are composed of smaller units of implementation, often called Configuration Items.

9.2.1 Configuration Items (CIs)

Configuration Items (CIs) are entities, below the Subsystem-level of the SoS architecture that must be developed and managed in order to deliver the SoS subsystems and elements. Configuration Items (CIs) represent elements of subsystems such as software services, hardware, firmware, network components, etc., that must be developed, implemented, tested, and delivered; these are managed through the SoS Configuration Management (CM) system. CIs include Service Level Agreements (SLAs) for SoS Enterprise services, operational sites (CIs can be buildings), The CI is the smallest unit within the SoS that can and will be changed independently of other components. CIs will vary widely in complexity and size, depending on the overall SoS design and implementation. CIs should be entities that are identified for maintenance in the operational system, and should be loosely coupled from other CI entities, being independently replaceable without affecting other entities, subsystems, or elements within the SoS architecture. This ranges from an entire service group (including hardware, software, networks, documentation, etc.) down to a single software service, hardware unit (e.g., workstation), or network device (e.g., router). This allows all issues, or DRs within the SoS to be isolated to one or more CIs so the changes can be made to resolve the deficiency. Each CI within a given subsystem is uniquely identified by identification code and version number. Often, software and hardware CIs are distinguished from other CIs within the system by titling them Computer Software Configuration Items (CSCIs) and Hardware Configuration Items (HWCI). If the SoS is primarily a software system, the term subsystem may be replaced by "Service Group."

Questions to think about:
What is the difference between service domains and services vs. the older Subsystem/CI vocabulary of this section, from an MDSE point-of-view?

9.2.2 Subsystem and Element Design Reviews

Figure 9.4 below illustrates the design and documentation flows required for SoS system and element design review. This includes documentation of all interfaces, CIs, software, hardware, systems, and testing designs for all CIs throughout the SoS. Whether separate documents, or paragraphs in an Agile Sprint Description, the information from these reviews must be under Configuration Management Control (even if it's just version control) to ensure each part of the SoS is in sync across the development efforts. The purpose of providing these process flows is for the MDSE to understand what needs to be taken into account when, throughout the SoS development life cycle.

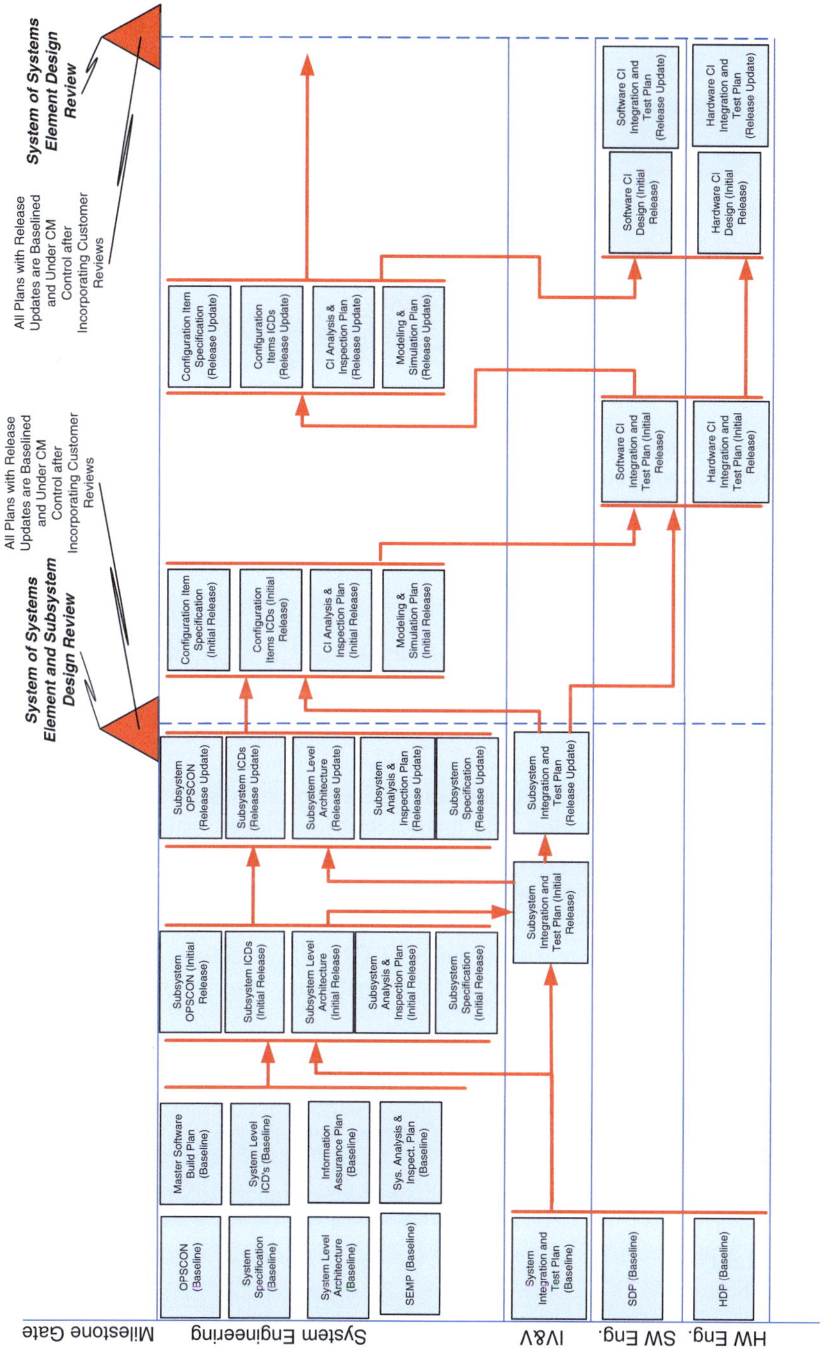

Fig. 9.4 Element and subsystem design reviews plan flows

Questions to think about:
What constitutes a Configuration Item?
Is it a group of divergent capabilities?
Is it more than 1 capability?
Is it a team organizational boundary?

9.3 Final Design Reviews to-Integration and Test Completion Process Plan Flows

9.3.1 *Final Design Reviews*

Many SoS programs will require prototype activities to verify the viability of certain capabilities required to meet the SoS requirements. These prototypes are utilized to drive the SoS final design. Once all CI, subsystem, and element designs have been verified and finalized, a complete Final Design Review (FDR) is initiated. This includes:

- Functional distribution across the elements, subsystems, and CIs
- Interaction between all elements, subsystems, and CIs required in order to provide SoS-level capabilities.
- Definition of all CIs (hardware, software, firmware, network, etc.)

Once the Final Design Reviews are completed, approved, and the system has been developed, testing can begin, including CI testing, subsystem integration testing, element integration testing, and finally, SoS integration testing. The FDR is required as a rigorous architectural and design review of SoS level capabilities and review of emergent behaviors which may have evolved, given the SoS element-element interactions. The FDR provides a formal review of not only overall system functionality against the system requirements, CONOPS, objectives, and constraints, but should also provide a review of the SoS capabilities in terms of properties that include, but are not limited to:

- Availability
- Flexibility
- Security (Information Assurance and Cybersecurity aspects)
- Reliability
- Interoperability (subsystem-subsystem, element-element, SoS to external interfaces, etc.)
- Adaptability
- Extensibility

At a high level, the FDR evaluates the SoS to ensure:

- The System and Software architectures and designs meet requirements.
- The required SoS capabilities are adequately represented.

Table 9.1 Final design review entrance criteria

FDR pre-condition	Potential documentation
Derived requirements satisfy and are adequately traced back to SoS/Element/Subsystem requirements baselines	Use Case and Activity Diagrams
Software increments/Sprints are planned and defined, consistent with requirements baseline	Software Build Plan, Sprint Plans, Sprint Backlogs, etc.
Hardware development is planned and defined consistent with requirements baseline	Hardware designs, specifications, and drawings
Systems/Software architectures established	Architecture plans/drawings, Class and Deployment (HW and SW) Diagrams
Requirements are satisfied by design approach	Sequence Diagrams
CI-level designs and specifications are complete	State Diagrams, Class Diagrams, Sequence Diagrams
Data Management is adequately specified	Data Architecture, Entity Relation Diagrams, Date Item Descriptions, Class Diagrams
Integration and Test plans are adequate	Use Case Diagrams, Use Case Specifications, and Sequence Diagrams

- All system constraints have been considered in the designs.
- Trade Studies exist for all major design decisions.
- High risk items have been addressed adequately (i.e., probability of occurrence reduces and/or consequences have been mitigated).
- Human Factors engineering has been adequately built into the design and the proposed SoS is usable.
- Operations and Sustainment has been adequately considered in the overall design and proposed implementation of the SoS.

The pre-conditions for holding the FDR (which, again, may be short if this is an agile development program) are outlined in Table 9.1.

9.3.2 CI (Service and Component) Integration and Test

Once each of the separate CIs (whether HW, SW, Firmware, etc.) are tested, each of CIs must be integrated together into operationally viable subsystems, then elements, and ultimately, the entire SoS. Integration of the CIs within a subsystem (CI-CI Integration) is essential to ensure the proper operation of the element subsystems before element-to-element integration. The overall architecture and design of the SoS should include the overall test philosophy and methodologies up front. Continuous integration of the system, both from the top down (SoS-Element-Subsystem-CI) and from bottoms up (CI-Subsystem-Element-SoS), is necessary to ensure a complete, operational systems. Utilizing Continuous Integration requires the MDSE to build into the architecture and design, reliable, secure, underlying enterprise services (e.g., data access services) and Enterprise Federation Policies in order to facilitate multi-level integration across the entire SoS on an ongoing and continuous bases. Continuous Integration is intended to be utilized with automated CI, subsystem, and element tests (called regression tests) to drive the systems to Test Driven Development (TDD) [118]. If the SoS elements are legacy systems, they may employ more than one Enterprise Application Integration technology that must be folded into the overall SoS integration strategy. Figure 9.5 illustrates the flow of information, from various SoS plans, through the Final Design Review (FDR), and through the initial CI integration testing.

9.4 Subsystem Integration and Test Process Flow

Subsystem Integration and Test follows CI testing (often called Unit Testing). The move to Subsystem Integration and Test activities cannot move forward until all CI testing and CI-CI integration testing has completed and any deficiencies found during CI testing have been resolved; all Deficiency Reports (DR) have been adequately resolved. During Subsystem Integration and Test activities, individual CI modules are combines to form subsystem groups. This continues through subsystem-to-element and element-to-System of Systems integration testing. Figure 9.6 illustrates an example that we will step through to illustrate subsystem and element integration and test [119].

In the example shown in Fig. 9.6, element A has four subsystems. Each subsystem is made up of two or more Configuration Items. From Fig. 9.6, we see that:

- Element A calls Subsystems A through D
- Subsystem A calls CIs A, B, and C
- Subsystem B calls CIs C and D
- Subsystem C calls CIs E, F, and G
- Subsystem D calls CIs G, H, and I

9.4.1 Top-Down Element and Subsystem Integration and Test

Utilizing a top-down approach requires lower level modules to be stubbed out in order the test the functionality and capabilities of the element or subsystem under test. From an element perspective, each subsystem is activated and must respond [120]. In order for the subsystem to respond, either the subsystem is stubbed out, or all of the lower-level CIs that are called by each subsystem, since, utilizing a top-down approach, the CIs haven't been integrated together yet. The integration and test process is continued until all subsystems are tested and functional; providing a test of the overall element architecture and design. This top-down approach allows major system threads to be exercised early in the integration and test process and functional problems addressed early on. This top-down approach can be integrated either with a horizontal (or breadth) approach, or a vertical (or depth) integration approach.

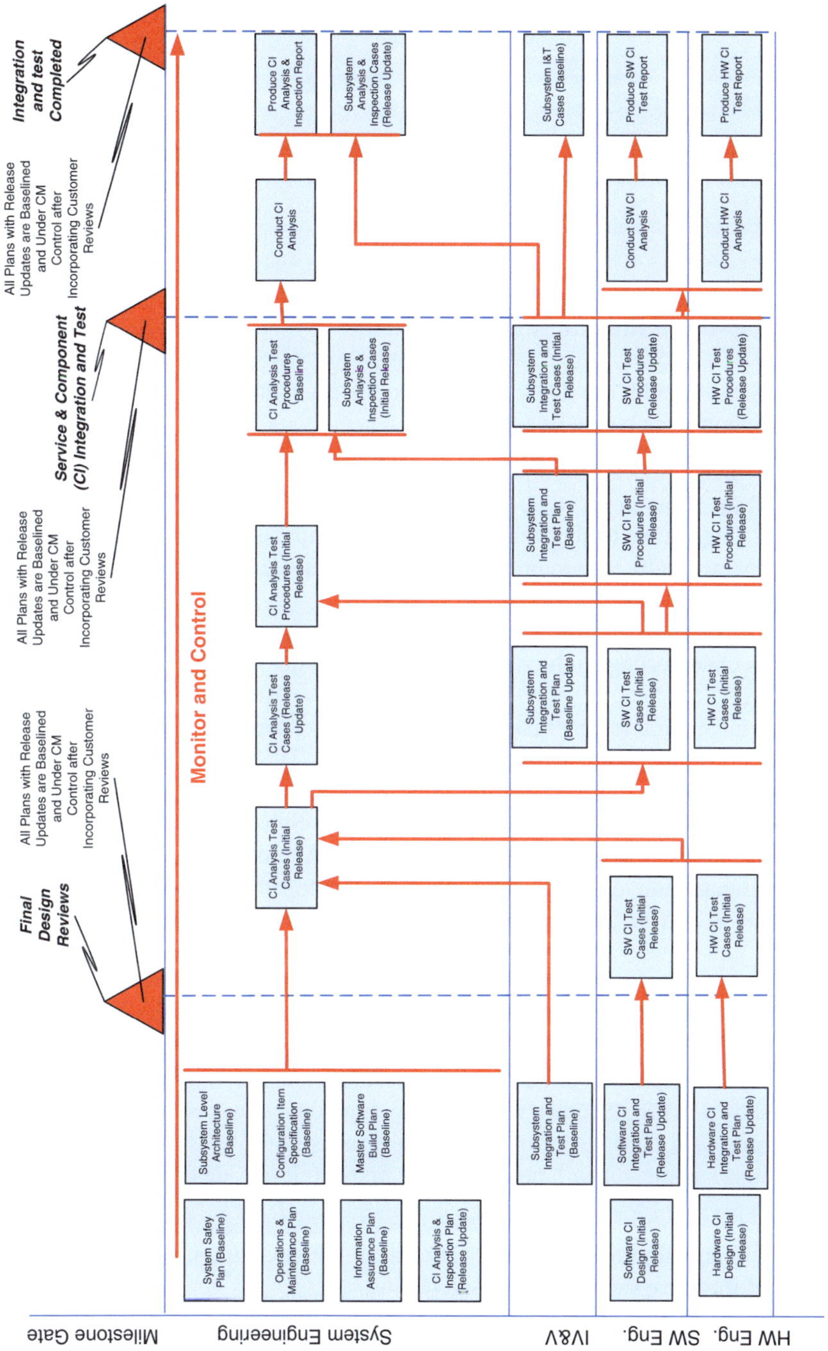

Fig. 9.5 Integration and test readiness flow

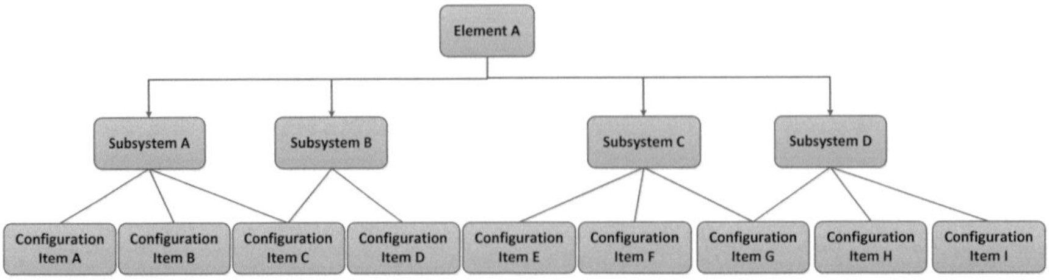

Fig. 9.6 Subsystem and element integration and test example

9.4.1.1 Top-Down Breadth Integration Testing

If we apply the breadth top-down approach to the example shown in Fig. 9.6, then the following process would be followed:

- Level 1 integration testing makes Element is the module under test; stubs are required for Subsystems A, B, C, and D.
- Level 2 integration testing makes Element A, and Subsystem A the modules under testing; stubs are required for Subsystems B, C, and D and for CIs A, B, and C.
- Level 3 integration testing makes Element A, Subsystem A, and Subsystem B the modules under testing; stubs are required for subsystems C and D and for CIs A, B, C, and D.
- Level 4 integration testing makes Element A, Subsystems A, B, and C the modules under testing; stubs are required for subsystem D; stubs are required for CIs A through G.
- Level 5 integration testing makes Element A, and Subsystems A through D the modules under testing; stubs are required for CIs A through I.
- Level 6 integration testing puts the entire element together, making the Element, Subsystems, and all CIs the modules under test.

9.4.1.2 Top-Down Depth Integration Testing

If we apply the depth top-down approach to integration testing approach to the example shown in Fig. 9.6, then testing takes on module down as far as testing can continue, and then the integration testing backtracks to the next major module to test. For this example you will notice that the first two levels are the same, and then it diverges.

- Level 1 integration testing makes Element is the module under test; stubs are required for Subsystems A, B, C, and D.
- Level 2 integration testing makes Element A, and Subsystem A the modules under testing; stubs are required for Subsystems B, C, and D and for CIs A, B, and C.

- Level 3 integration testing makes Element A, Subsystem A, and CI A modules under testing; stubs are required for Subsystems B, C, and D, and for CIs B and C.
- Level 4 integration testing makes Element A, Subsystem A, and CIs A and B modules under testing; stubs are required for Subsystems B, C, and D, and for CI C.
- Level 5 integration testing makes Element A, Subsystem A, and CIs A, B, and C modules under test; stubs are required for Subsystems B, C, and D.
- Level 6 integration testing makes Element A, Subsystem A and B and CIs A, B, and C modules under test; stubs are required for Subsystems C and D, and for CI D.
- Level 7 integration testing makes Element A, Subsystem A, B, and C and CIs A through D modules under test; stubs are required for Subsystem D and for CIs E, F, and G.

This continues until all the Subsystems and CIs have been tested top-down. Next we will explore Bottom-Up integration for the same example.

9.4.2 *Bottom-Up Element and Subsystem Integration and Test*

As the name implies, with Bottom-Up integration testing, the CIs are exercised and tested first. After testing and integrating the CIs for one subsystem, then the CIs for the next subsystem, etc., until all subsystems have been tested, then subsystem integration testing begins. For our example shown in Fig. 9.6, integration testing looks like:

- Level 1 integration testing makes CIs A, B, the modules under test; stubs are required for CI C and internal and any external interfaces between subsystems that involve Subsystem A.
- Level 2 integration testing makes CIs A, B, and C the modules under test; stubs are required for any external interfaces between subsystems that involve Subsystem A.
- Level 3 integration testing makes CIs C, and D the modules under test; stubs are required for any internal and external interfaces between subsystems that involve Subsystem B.
- Level 4 integration testing makes CIs E and F the modules under test; stubs are required for CI G and for any internal and external interfaces between subsystems that involve Subsystem B.
- Level 5 integration testing makes CIs E, F, and G the modules under test; stubs are required for any internal and external interfaces between subsystems that involve Subsystem B.
- Level 6 integration testing makes CIs G and H the modules under test; stubs are required for any internal and external interfaces the involve Subsystem C.
- Level 7 integration testing makes CIs G, H, and I the modules under test; stubs are required for any internal and external interfaces the involve Subsystem D.
- Level 8 integration testing makes Subsystems A and B the modules under test; stubs are required for Subsystems C and D and for any internal and external interfaces that involve Element A.

- Level 9 integration testing makes Subsystems A, B, and C the modules under test; stubs are required for Subsystem D and for any internal and external interfaces that involve Element A
- Level 10 integration testing makes Subsystems A, B, C, and D the modules under test; stubs are required for any internal and external interfaces that involve Element A.

9.4.3 Integration and Test Process Flow Through Subsystem Integration and Testing

Figure 9.7 below illustrates the Integration and Testing plan and process flow through completion of Subsystem Integration and Test.

9.5 Systems Integration and Test Development

The final top-level System of Systems integration and test activities follow the same procedures as discussed in Sect. 9.4, depending on the methodology chosen. Figure 9.8 illustrates this plan and process flow through final System of Systems testing, bringing the system to the point where it is ready to ship to the customer operational site (which may be internal to your company.

9.6 Alternative Integration and Test Methodology: Functional Testing

Most programs utilize either top-down or bottom-up integration and test strategies (although bottom-up is more prevalent). However, it is possible to utilize a hybrid approach to try to attain the advantages of both methods, while trying to avoid the disadvantages of both, called Function Integration and Test. In Functional Integration and Test, capabilities required by the System of Systems are collected into functional groups (logical subsystems), rather than testing per the Work Breakdown Structure, as many programs do. This type of Integration and Test methodology requires the MDSE to design this methodology into the architecture, design, and implementation throughout the program life cycle. The Functional Testing methodology may employ a combination of top-down and bottom-up integration testing, depending on the grouping of functions within the System of Systems context. Some potential functional groupings that are possible, depending on the System of System capabilities are:

- Interface Modules for each data store interface (data services layer).
- External I/O interfaces
- Command, or transaction, based logical subsystems. If the SoS were a banking system, this might involve logical grouping of the debit and credit transaction software for testing.

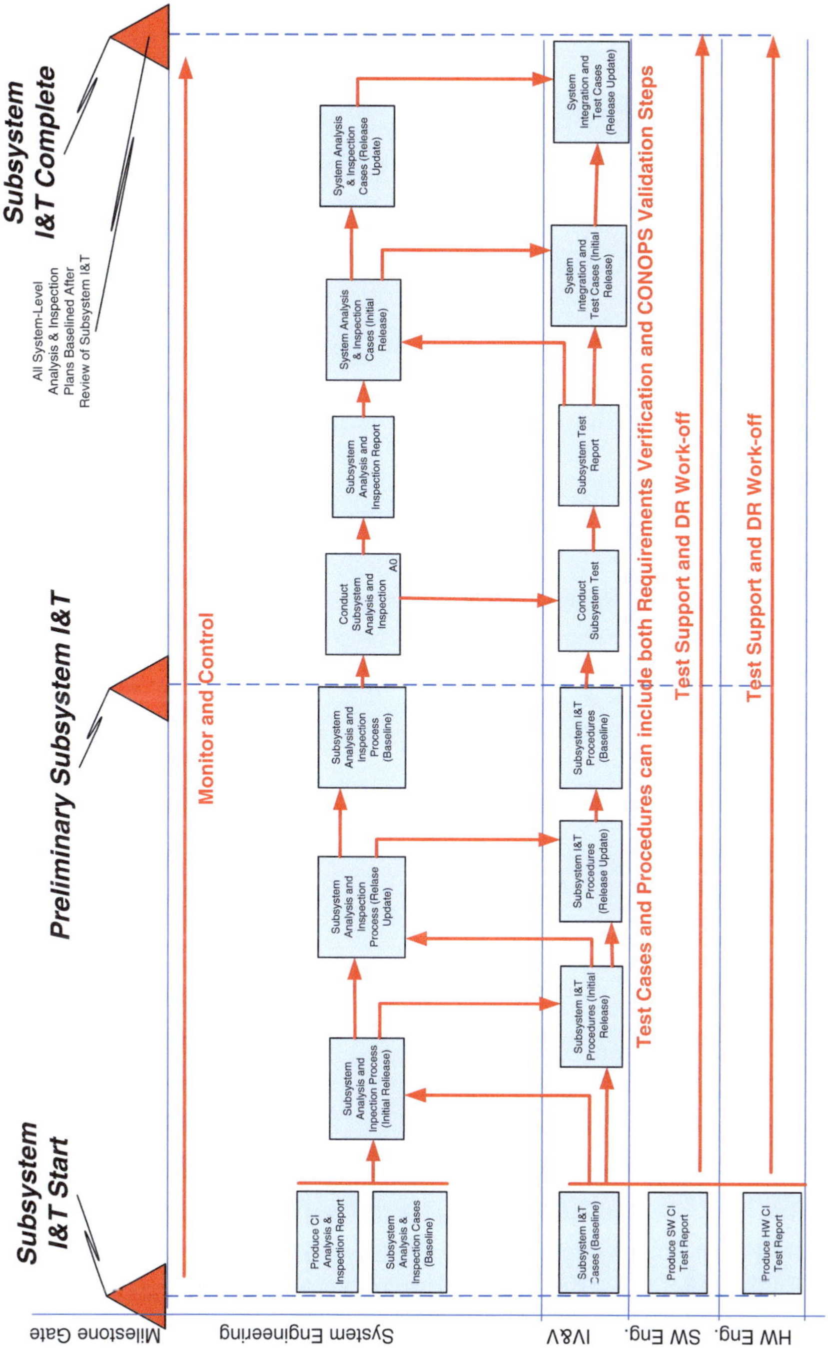

Fig. 9.7 Subsystem integration and test flows

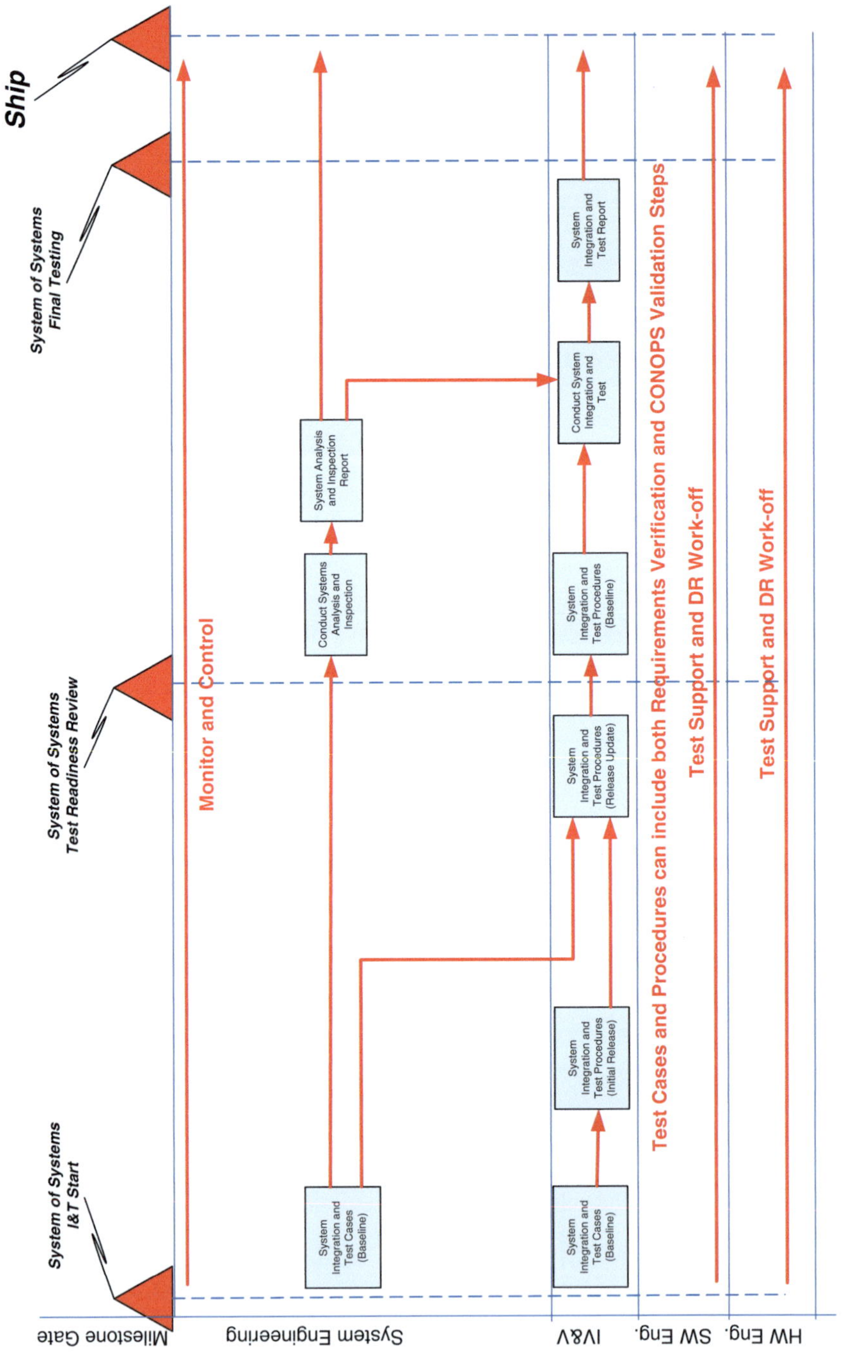

Fig. 9.8 System of systems integration and test flows

- Computer security-related services and applications (i.e., identification, authorization, etc.)
- Data Access Services

9.7 Case Study: What Happens When Test-Driven Design Isn‘t: The Problem of Test Data

9.7.1 Garbage in Garbage Out

The quality of the SoS architecture depends, to a large part, on the inputs provided to the MDSE architect. The MDSE architect is responsible for ensuring high-quality inputs into the SoS. For example, the MDSE should identify sources for the SoS User's needs appropriately, obtaining User's needs from the Users, not from speculation by those who have never used a system similar to the SoS being developed [121]. If low-quality, noisy, inaccurate, or low-quality upstream information is provided to the MDSE, the system will result in a sub-par architecture. Communication, cross-checking, and other information gathering and verification approaches can, and should be, used to ascertain the quality of information being used to design the system architecture. This is particularly true in system test and integration. The ability to state an idea simply does not necessarily or frequently lead to simplicity in the execution of the idea. The architectural concept or design of testing procedures may seem simple or logically flow on paper, but without the proper data required to operate the system and ensure that the data represents that which will drive the system in ways consistent with the tests required to ensure it meets its requirements, real-world testing will prove problematic at best. An issue with integration and testing of SoS elements, subsystems, and CIs, is making sure all development groups have the same idea what testing means and the data and test scenarios required for an overall integration test.

9.7.2 Providing Data Management Across the SoS

Unfortunately, required test data is often neglected until the integration and test events are close at hand. Data, especially data required to test a system, is the lifeblood of successful system development and deployment. Test data must be architected into the entire process to ensure it is accessible, accurate, complete, and appropriate for the required testing. It is essential that the MDSE recognize the overall need for Enterprise Data Management (EDM) and its effect on the SoS development essentials:

- **Mission/Business Data Content Management**: the display and dissemination of data/information throughout the SoS, including the integration and presentation of unstructured data produced throughout the elements, subsystems, and CIs.

Table 9.2 SoS data management attributes

Data management attribute	Description
Data architecture	The data flows, data access services, processes, software (COTS—like Oracle) and procedures required to access, store, retrieve, and operator responsibilities for the data at each tier of the SoS
Data structures	How data is organized (e.g., database schemas, flat file structures) across all tiers in the architecture
Metadata	Information that is captured about data to provide the data context. Metadata increased the interoperability of elements and subsystems
Data governance	Provides the procedures, policies, roles, responsibilities, and structures that guide data management across all of the SoS data
Data security	Provides policies, procedures, software, firmware, and potential hardware (e.g., data guards) to protect data/information from unauthorized access, viewing, modification, creation, deletion, including metadata, whether unintentionally or intentionally
Data accuracy/quality	Defines the requirements for the accuracy and quality the data is required to have/attain across all tiers of the SoS. This includes data completeness, legal compliances, including external user data needs

- **Situational Awareness (Mission/Business Intelligence)**: How data is combined, correlated, analyzed, and reported to support decision making across the SoS users and user community.
- **Mission/Business Processes**: the internal and external processes (manual and automated) that utilize and depend on data throughout the operational life cycle of the system (from initial delivery of the system until retirement of the systems).
- **SoS Enterprise Data Services**: those services throughout the SoS that provide data/information on demand to meet specific mission/business needs, test scenarios, and training; any and all data needs. Data services should ensure that the data provided across the SoS is unadulterated (clean), current, complete, accurate, relevant, and secure.

Table 9.2 below outlines the attributes of a successful Data Management Strategy that the MDSE must ensure is built into the SoS architecture, design, processes, procedures, and activities throughout the SoS development and operations life cycle.

9.7.3 Test-Driven Design and Test Case Development

Test-Driven Design (TDD) involves development of test cases and test philosophies before software is written. TDD drives the software developers to work with the systems engineers to think through the requirements and develop methodologies and structures for testing the software up front, at the beginning of the project, and not as an afterthought when the project is in a panic to get through integration and test when the project is behind schedule (which happens more often than not). TDD is especially important in agile design/development methods.

Often, in classical integration and test, code is written, and then tests are created to verify and validate the code. Correct practice is to have an independent group develop the test cases/scenarios based on the requirements the software is supposed to be written to satisfy. When projects get behind and test cases are left to the end of the software development cycle, it is commonplace to have the software engineers that wrote the code, write the test cases to test their code. This allows the potential for limited test to be created in such a way that the tests do not fail. If they fail, then the code is updated just enough to pass the part that failed. This rarely results in test cases that test all of the possibilities, including error conditions that the software requirements dictate.

The basics of TDD boil down to:

- **Test with a purpose**: know why the code is being written and why it is being tested before a single line of software is written.
- **Granularity of the testing should be based on the importance of the system**: a toaster should be tested differently than an airliner control system.
- **Test 100 % of the code**: One of the side effects of TDD is that you cover 100 % of the functionality the software is written to achieve, since the functionality is built into the test designs up-front. Traditional testing doesn't guarantee complete coverage and rarely achieves it.
- **Provide documentation your software developers will actually use**: rarely do developers read all the documentation provided to them. Find out what documentation they prefer to use and make sure that documentation is complete enough to specify the requirements, functionality, constraints, etc., for the software.

9.7.4 Integration Testing Data

Integration Testing Data is data created ahead of time, generated at test time, or recorded from live operations, for the specific use of in CI-level, subsystem-level, element-level, or SoS-level testing. Some test data is utilized to verify and validate overall system functionality, while other test data is used to understand how the SoS responds to errors (anomalies), user mistakes, or corrupt data; any type of anomalous, exceptional, incomplete, corrupted, or unexpected inputs. Testing may involve focused scenario required to test specific functionality (or system threads) required of the overall SoS, or can be high-volume testing that is stochastically generated and may involve automated testing [122].

One of the major difficulties in software testing is the creation of useable test data. This is very labor intensive and may account for up to half of the efforts involved in Verification and Validation. Test data generation is a complex problem and those providing easy solutions can only do so for very small, focused systems. Test data generation for complex software, especially in a SoS is problematic since it is hard to anticipate the paths the software will take during the execution of test data, making the outcome highly unpredictable [123]. Also, it is difficult for Test Data Generators (simulators) to produce exhaustive amounts of test data that cover a variety of scenarios and test cases. Failure to provide the quality and quantity of test data required to thoroughly test a SoS can be disastrous when the system is put into operations.

Some of the issues involved in data creation, preparation and use in SoS integration tests are outlined below [123]:

- Preparing test data that will exercise system requirements in a scaled-down non-production hardware environment.
- Managing concurrent sets of test data requirements in multiple test environments (element-element and subsystem-subsystem).
- Ensuring adequate and appropriate test data exists for all applications in the integration test environment.
- Inability to utilize actual production data for integration testing.
- Ensuring Information Assurance policies and procedures within the test environment to prevent exposure of sensitive data.

9.8 Discussion

In the end, the MDSE systems engineering process is an iterative problem solving exercise to create an architecture, design, and operational system. The problem solving activities involve analysis, evaluation, and synthesis of all of the requirements, goals, objectives, and constraints posed for the proposed SoS. The MDSE team's responsibilities are to provide and manage the processes, procedures, and artifacts throughout the SoS life cycle. The MDSE approach is far different from classical systems engineering. Figure 9.9 illustrates a classical mind set for systems engineering.

In classical Systems Engineering, the contributions provided by the Systems Engineering organization decreases as the SoS development life cycle progresses. In contrast, MDSE is majorly involved in all aspects of SoS design, development, integration and test, and deployment. The overall MDSE process is illustrated in Fig. 9.10.

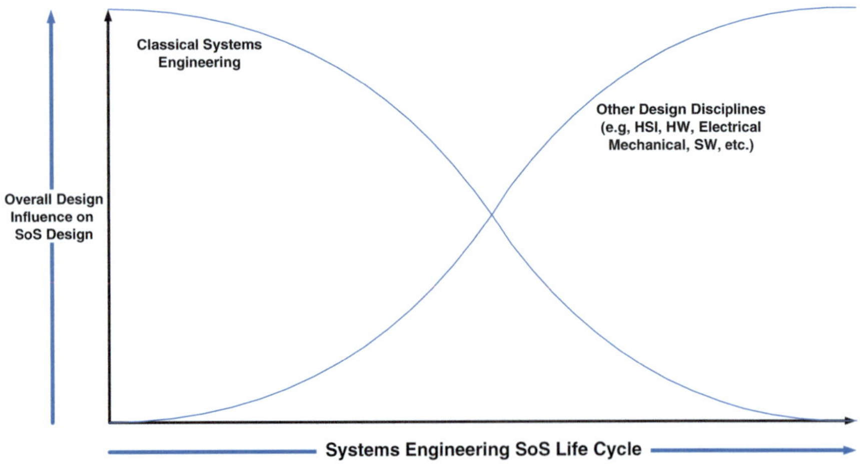

Fig. 9.9 Classical systems engineering influence on the SoS design and development process

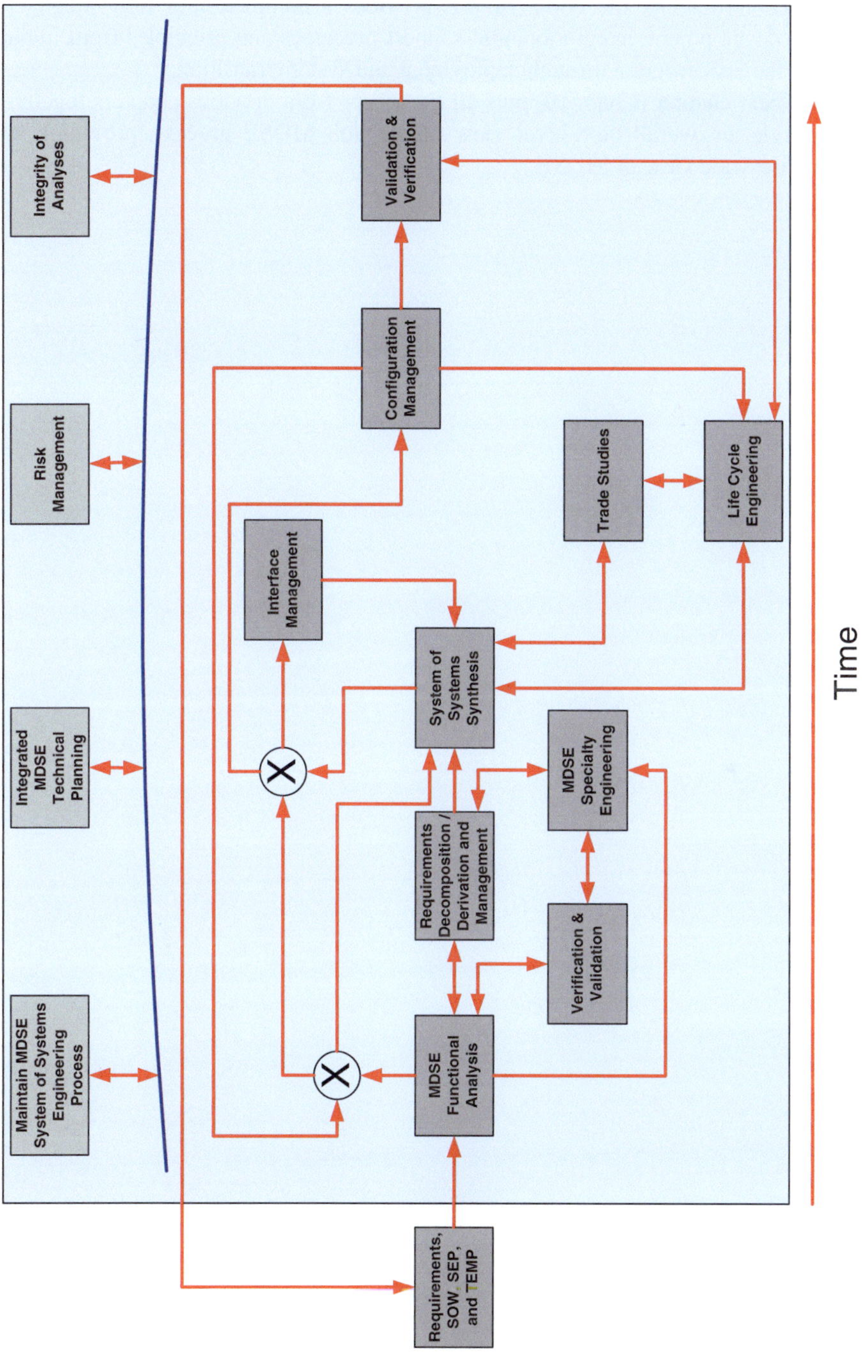

Fig. 9.10 MDSE influence on the SoS design and development process

MDSE is a continuous engineering process involving multiple disciplines as has been discussed throughout the book. MDSE provides a complete and well-integrated design, development, integration, and support processes and principles from initial start of the SoS program through deployment and system transition.

The next chapter (Chap. 10) puts all the pieces from previous chapters together to provide an overall high-level view of the SoS MDSE process, providing an enterprise-wide view of MDSE.

Chapter 10
Putting It All Together: System of Systems Multidisciplinary Engineering

Developing relatively small solutions, to very large Systems of Systems, requires today's Systems Engineers address multiple disciplines in their daily duties. Without realizing it, today's systems, and the systems of tomorrow, require a Multidisciplinary Systems Engineering (MDSE) solution. The background and materials in this book are an essential step to implement a change from the stove pipes engineer roles of today, into a well-rounded Multidisciplinary engineering role of tomorrow [124].

10.1 Taking the Enterprise View of Systems Engineering

When developing a new system, the MDSE engineer must approach the solution from an enterprise view. This means that they must evaluate the objectives of the customer (or corporate management team) in relation to the capabilities being procured and developed, and how it replaces or integrates into the enterprise.

The MDSE engineer starts the solution at the top of the enterprise, and moves systematically from layer-to-layer of the system to the lowest capabilities being development relative to their perspective of the system. By following the top-down approach, and addressing the multidisciplinary issues at each tier of the development, the solution can be evaluate and assess in relation to its objectives, requirements, interfaces and operational usages.

10.1.1 Perspective

Depending on the size and scope of the development, the MDSE is faced with one or more tiers of system refinement. A large system may come with multiple tiers depending on the departure point to development, whereas, a system of systems may have many more tiers. Based on the perspective, the new development may be a component of a system of systems, a system, segment, element, etc.

J.A. Crowder et al., *Multidisciplinary Systems Engineering*,
DOI 10.1007/978-3-319-22398-8_10

The following example provides perspective for both enterprise and system of systems [125].

(a) The United States Department of Defense (DoD) is a component of a enterprise (e.g. United States Government)
(b) The DoD has multiple branches (e.g. USAF, USN…)
(c) USAF branch may have multiple systems
(d) …

If the US Executive branch is the customer (as authorized by the US Congress), the point of departure for the development may be a new branch of government (e.g. Homeland Security after 911).

If the USAF needs a new space system, the new system is much lower in the USG Enterprise, than Homeland Security.

The point being made is: The enterprise view to the system is defined by the perspective of the customer/management in relation to the size and scope of a development.

The following example will take a segment of a system, and decompose a leg of the ground segment to an individual capability to expose the tiers of refinement. The *italic* items in the thread being decomposed are part of a control station's N-tier architecture, where the N-tier architecture is contained in a processing enclave.

Example of a Space System:

(1) System—a system which has multiple segments (e.g. space, ground, users)
(2) Segment—Space, *Ground*, Users
(3) Element—*Control Station*(*s*), Ground Antenna(s)
(4) Sub-Element—Enclaves (*Operational Center*, DMZ(s), Server/Data Farm(s)…)
(5) Service Domains (e.g. Sub-Systems)—*Presentation* Tier, Business Tier…
(6) Service Tier (e.g. CI)—*Display* Services
(7) Component—*Global Map* services
(8) Unit—capabilities on the map (e.g. space assets, ground assets, *Ma*ps, User commanding interface…)
(9) Capability—Digital Terrain Elevation Data (DTED) map overlay

As you can see, there are multiple tiers of specification, architecture and like items that must be assessment for a large system. A development may be allocated to any portion of the system, but that development should not begin until the objectives and requirements are established for the capability being developed.

In the Space System example above:

(a) A new Space System may be the development (as a component of a system of systems)
(b) a new Satellite in the space segment may be the development
(c) a new processing enclave may be added to a control station as the development
(d) …

The purpose of the examples above is to expose both perspective to enterprises, system of systems and a system within an enterprise. By following top-down engineering, the MDSE engineer will develop a solution for each tier of the system until the level of refinement for the perspective is satisfied. In the top down effort, the principle tier will consume the customer/management's procurement/development objectives, constraints, requirements, interfaces as well as operational concepts. As discussed in other chapters on requirements and architecture and tools, the MDSE flow will take the primary inputs and begin the refinement cycle. A sound cycle is:

Step 1: Establish the tier's requirements—derived from CONOPS, Objectives, Requirements, Interfaces and alike
Step 2: Perform an architectural analysis of the tier
Step 3: Design the operational concepts and/or Operational Views (OV) of the tier
Step 4: Design the system concepts and/or System Views (SV) of the tier
Step 5: Assess the tier's solution

Based on each tier, the requirements, architecture and like artifacts should be defined at the level of fidelity required for system understanding.

Back to the Space System example, the system requirements at the first Tier should define the primary segments, and each segments principle capabilities required.

Step 1: A good system requirement for a segment would be:

The ground segment shall provide the capabilities to manage on-orbit operations of the satellite constellation as defined in the list of on-orbit satellites….

The requirement is broad and provides the capability the ground segment needs to provide, with bounding to the space segments list of satellites.

Step 1: A poor system requirement for a segment would be:

The ground segment shall provide the capabilities to command the thrusters on a satellite in accordance with the satellite interface control document(s).

While the requirement is valid, the requirement is poorly constructed for segment definition, as it provide for the Segment to worry about a specific lower level component on an SV.

Step 2: A good architectural assignment/allocation of the segment requirement would be:

The control station provides SV telemetry, tracking and control functions
The ground antenna(s) provide the ground to space communication functions

Step 2: A poor architectural assignment/allocation of the segment requirement would be:

The control station provides bus commanding
The ground antenna(s) provides uplink operations on s-band frequency 1

This example provides the opportunity to expose what does actually happens in the real world, where the actual requirements sets from the customer comes with these varying levels of fidelity. It is the responsibility for the MDSE to level the specifications. This requires extensive analysis of the baseline materials provided by the customer/management teams. Once the specifications, and architecture analysis has been leveled to the appropriate tier fidelity, it's time to complete the cycle for the tier.

Each tier is then decomposed along functional boundaries as established by the higher level architectural assignment. In our example, the control segment provides for the operators to management the satellites, whereas the ground antennas provide the electronic communication links (typically RF) to space.

A general rule to follow is: The fidelity defined in each tier, and the number of tiers, can be negotiated in relation to cost and schedule objectives, as long as the objectives for the system's development and long term support can be achieved.

By taking the enterprise view to the system, and developing the system top down, may lead to more changes than originally envisioned by the customer/management team, but by factoring in their evaluation and feedback, the top-down approach allows systematic approach to change, rather than the de facto changes that occur when the new development is found to be defective.

Top-down assessment, with continual feedback, typically closes on the objectives more often than naught. Whereas, bottom-up changes typically fail to achieve the system objectives. The next section address the pit falls of bottoms-up engineer. Before we go there, the topic of reuse must be addressed.

As we have stated many time throughout this book, most systems are enhancements to existing systems, or where the developments reuse existing systems to build new systems. The top-down engineering approach must treat reuse products as an opportunity, and not an end-all solution for the development. While true, the new system may be similar to the existing reuse, the purpose of a new development typically comes with new requirements that older systems in the field do not properly address.

When building a new system with the intent to reuse existing solutions, the top-down enterprise view requires the MDSE engineer ignore the reuse solution space as the end all answer. They must properly decompose the customer/management objectives, requirements, concepts of operations, interfaces and alike. When the primary tier's requirements are established, the architectural analysis should identify where reuse supports the requirements, what modifications are required to the reuse, and lastly, what new capabilities must be built to satisfy the procurement/development. This allows the tier to be designed to completion, and start the next tier.

For a solution to be considered reuse it must have certain characteristics, these characteristics are very easy to understand, especially in relation to DoDAF materials. Companies with valid reuse libraries typically have metadata catalogs for the capabilities, which typically provide:

(1) A description of the capability
(2) Requirements for the capability

(3) Architectural pattern for the capability
(4) OPSCON/Operator Use Cases/OV views
(5) UML designs, tested code/SV views
(6) Test cases: automated, procedural or both

There may be other artifacts that support the reuse, but the primary items above provide proper definition of reuse. If the reuse is a code library without the support materials noted, then the code is just widgets and not true reuse. True reuse should come with the artifacts that allow the MDSE engineer to select and perform a full analysis on the capability, update the requirements, and alike artifacts satisfy the tier the capability is being applied to.

10.2 The Pitfalls of Bottoms-Up Engineering

The developments that incorporate change by starting with a base system, and then trying to adapt it to fit requirements, typically fail to meet the objectives of the new procurement/development. While most engineering teams believe that they have the answers to the issues, most teams focus on how they will change the existing system, or build the system at the lowest tiers and work their way back to the objectives [126].

This is classical arrogance we find in many different and divergent organizations. The personnel have develop a system solution, that when adapted, will satisfy most (where most) objectives and requirements. It is the "where most" problems start. When bottoms-up engineering is used in a development, important objectives, requirements and solutions typically fail to be realized. The size and scope of the pitfalls vary, but in general, the customer's and/or management's realization of the new and improved system can come at higher costs, extended schedules, and in some cases, may lead to project cancellation.

Most all bottoms-up development fail one or more objectives, where the objectives vary from:

(1) Procure on a fix cost profile
(2) Procure on a fix schedule profile
(3) Insert a new capability without radical change in functionality
(4) Time to Market
(5) Enhance performance, security, data, processing, and alike capabilities
(6) Sustainability
(7) Operability
(8) LCC
(9) …

There are a number of reasons that bottoms up will fail to satisfy objectives or requirements. The failure signatures will vary, and failure occurs more often than naught. The size of the failure and loss of satisfaction will vary based on overall system objectives. Using the N-Tier architectural model for a system solution, some of the typical problems found in bottoms-up solutions are:

(1) Human-System Interfaces (HSI) lack conformity to HSI design standards
(2) Data Management problems
(3) Interface Complexity
(4) Security Management problems
(5) Redundancy in code
(6) Increased integration and test cost
(7) Documentation shortfalls
(8) Extensive LCC cost issues

10.2.1 HCI Failures

Typical in reuse systems, they come with existing Human-Computer Interfaces. The displays were built on technologies that are older, or worse, obsolete. To save money, the developers will typically ignore the customers HCI standards, until it's too late, or worse, decide that having multiple different looks and feel solution are acceptable to save money. This always bites the customers in the backside, when their operators have to have multiple training programs to understand how to use the different reuse products.

Example: A system delivered to a USG organization with three different display solutions for a single operator (e.g. displays build on Windows, JAVA, HPOv) all with different Color, Fonts and alike issues.

Result: Customer dissatisfaction—lost profits—next maintenance upgrade came with specific requirements to clean up prior mess.

10.2.2 Data Management Problems

Whether or not the system development is using reuse, when data management is performed by many teams without a top down solution, they run head-long into problems. The databases and flat files designs will vary team to team driving up the costs of storage/retrieval, as well as archive/retrieval. Typical shortfall will include problems in data labeling, data records management, data version controls, Mission Assurance (MA), and Information Assurance (IA) problems. This one issue is why most developments get stopped and repaired or abandoned. Bottoms-up data management should never occur.

Example: A system built for an USG organization with 14 different storage solutions (e.g. Oracle DBMs, XML files, Binary Files, ASCII files…), without data normalization and labeling for MA or IA.

Result: Contract Termination

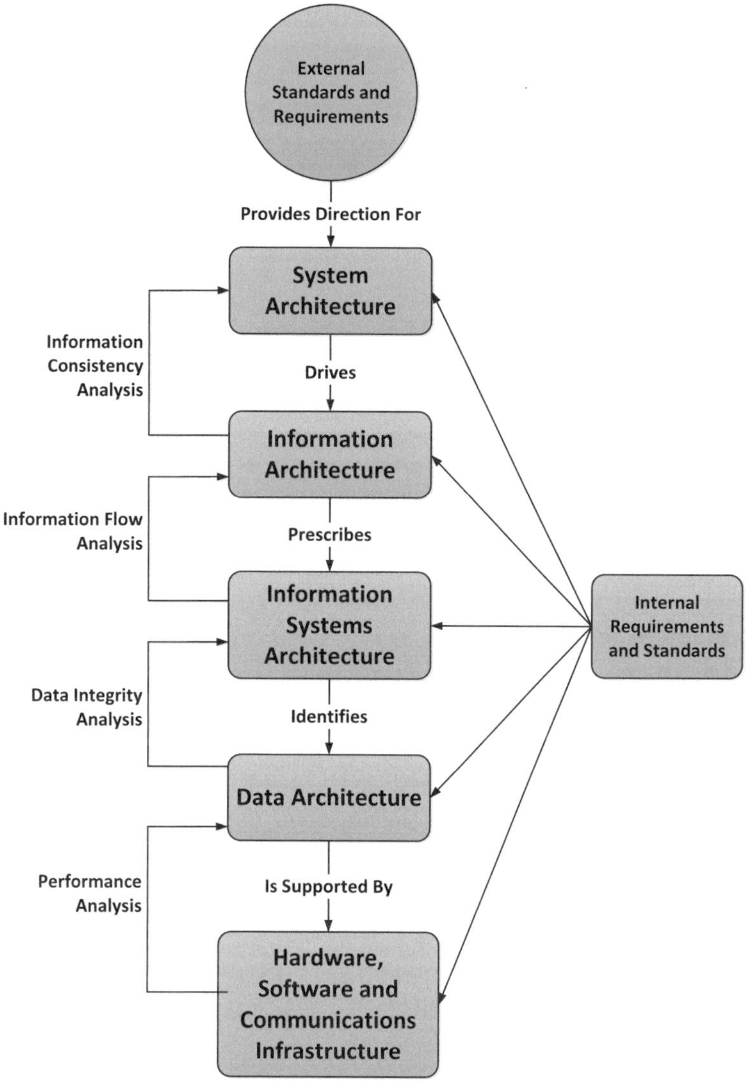

Fig. 10.1 The MDSE SoS data architecture process

Figure 10.1 illustrates the process of creating an MDSE Data Architecture process, based on a multidisciplinary approach that flows systems, data, information, hardware, software, and communications disciplines together to create a formal SoS Data Architecture. As with all MDSE approaches, creation of the Data Architecture is a feedback-driven that encompasses system performance, integrity, information flow, and data consistency across the entire SoS Enterprise to ensure the Data Architecture supports the overall system.

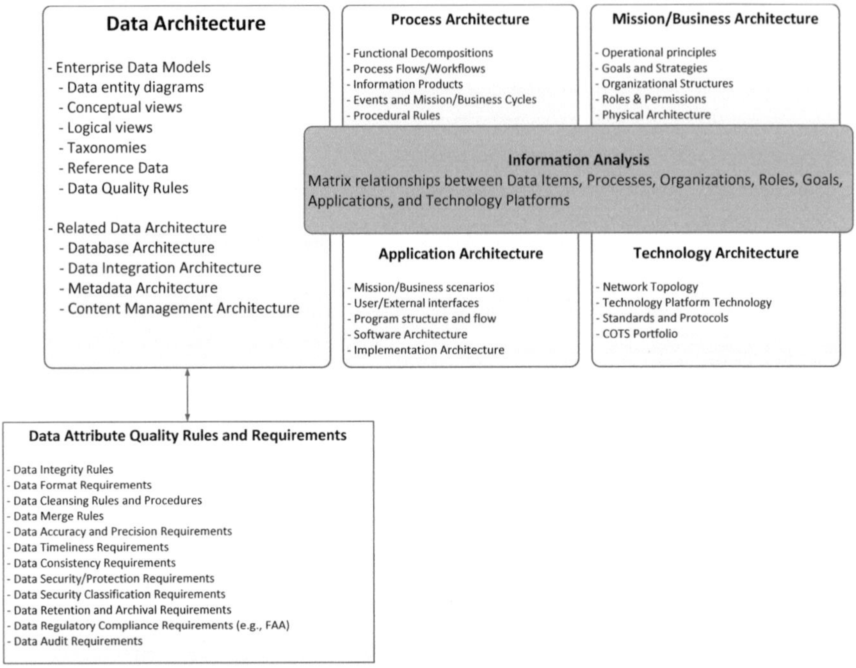

Fig. 10.2 MDSE SoS data management process

Once the Data Architecture is designed, it must be managed throughout the development, integration and test, deployment, operations, and maintenance; throughout the entire SoS life cycle. Figure 10.2 illustrates the Data Management Process.

The process illustrated in Fig. 10.2 is crucial (as discussed above) to ensure the Element-Element and Subsystem-Subsystem Enterprise data/information flows throughout the SoS. Figure 10.2 identifies all of the data architecture views, rules, requirements, and processes the must be implemented (at some level) in order to have an effective Data Architecture.

10.2.3 Interface Complexity

This is a broad subject, as this includes inter-system interfaces, intra-system interfaces, interfaces between COTS products, interfaces between developed code and COTS products, interfaces between developed code and developed code, software to hardware interfaces and alike across all interfaces. The ability to bridge and/or isolate capabilities via interfaces should be developed top-down. Bottoms-up interfaces tend to fall apart the minute multiple development teams are working on different solutions that need to interface to each other to form a service solution. Teams invariably use different versions of the interface schema, service definition, protocols, etc. This leads to delays in integration and rework, all leading to cost and schedule overruns.

Example: A hardware system required extensive rework, when the hardware/electrical interfaces didn't integrate because the mechanical 3 pin connector didn't mate with the 4 pins connector.

Result: Customer dissatisfaction, Cost/Schedule growth—lost profits

Example: A system under development trying to integrate multiple services, where the services had overlapping port/protocol assignments.

Result: Customer dissatisfaction, Cost/Schedule growth—lost profits

10.2.4 Security Management Problems

This occurs on both Data and Interfaces. Bottoms up solutions typically fail to incorporate the required header information need for the security firewall and/or guard protections. Bottoms up interfaces typically have ports and protocol issues affecting network security solutions for the interfaces, which includes multiple capabilities all trying to use the same or overlapping ports and protocol.

Example: System data provided no security labels on data. Firewall/Guard rules had to become code like to read the data to execute data blocking rules.

Result: multi-year schedule slip, extensive cost overruns—contract termination

10.2.5 Redundancy in Code

Code redundancy across teams is a typically failure in bottoms-up engineering. The teams will develop duplicative code to ensure their specific capabilities come together, resulting in like code across multiple code files used to build the system. This drives the cost of discrepancy fixes up, integration and test costs higher, and worse, the sustainment cost of the system escalates as well.

Example: A system under development for USG, went to stop work due to cost/schedule delays, and system performance issues. Tiger teams resolved system requirement and concept of operation issues, and operational prototypes developed. During prototype development 160K software lines of code removed from the code base improving system reliability.

Result: Contract Restart, multi-year cost/schedule delays reoccurred cause by bottoms up development, leading to contract termination 9 years after initial acquisition.

10.2.6 Integration and Test Costs

This category of costs always increases in bottoms-up development, and will result in large cost increases and schedule delays. The prior five factors (HCI, Data, Interface complexity, security and redundancy in code) all come together to make integration of the system or system of systems difficult (mobile failures), time consuming (DR fixes), misalignment of interface and methods for the interfaces…

Example: A system under development entered into integration phase without formal controls over the interfaces (software to software, software to COTS…). Three (3) years to integrate initial delivery.

Result: Customer dissatisfaction, extensive Cost/Schedule growth—lost profits w/ cost sharing penalties.

10.2.7 Documentation Shortfalls

Because the system is being built bottoms-up, the documentation for system, software, data and hardware in the solution space will typically be misaligned, and potentially developed with different products (e.g. Office Products, Case Tools, Drawing packages…) depending on how large the developer base is. This will occur as each team tries to document the system as the capabilities are building up. Trying to get a single data package built to provide to the sustainment team becomes next to impossible. It is always better to define the Documentation Products to be used by all teams Top to Bottom, to include formal Data Item Descriptors. When this does not happen, then each team may interpret the documentation solution.

Example: New system under development. System documentation provide by the customer for development teams found to be overcome by events during legacy system sustainment. New system off-course in terms of requirements, concepts of operations and architectural improvements required [127].

Result: Customer accepted development gap caused by missing documentation, initiated technical recovery actions, provided for increase in Cost/Schedule growth, development team…LUCKY!

10.2.8 LCC Growth

Bottoms-up development, if they survive the prior types of failures, will result in Life Cycle Cost growth. The system maintainers will potentially have specialized training to support multiple operating systems, databases file structures, documentation suites and alike. LCC must be a primary objective flowed top to bottom [128].

Example: New development transition to operations. Sustainment team provided with minimal documentation, and insufficient spares.

Result: Customer dissatisfaction, system restricted from operational usage until documentation for operations and maintenance met minimum standards, and sufficient spares provide to meet operational sustainment requirements. Delay is system use for 9+ months.

10.2.9 Reactionary Engineering

The behavioral aspect of bottoms-up engineering can be referred to as reactionary engineering. Reactionary engineering occurs as the system development begins to fail, and the engineering teams, then management, then customer begin to react to the bad news and progress. Without a clear top-down solution (road map) being maintained, changes in the development to overcome pitfalls leads to unplanned failures in other areas, and eventually a cascading effect occurs, where the fixes and counter fixes leads to complete technical breakdown. In the early stages of reactionary engineer, and with little capital expended, the management and customer teams will provide more budget and schedule, and authorize risk mitigation activities. Eventually, bottoms-up technical merit will continue to erode to the point, where low level capabilities are removed from the plan, then medium and then high need items. When the shedding of capabilities or the quality of the engineering gets to the pain threshold, chaos begins to happen within the customer, management and engineering departments.

The mitigation plans begin to fail, leading to more oversight, more mitigation plans, more support, more oversight, etc. The engineering teams will literally be moving backwards in progress, and in some cases will have introduced too many bad corrective actions such that the system will never integrate. Eventually the program will be terminated.

It is exceptionally important for the customer and management teams to detect bottoms-up engineering and terminate the practice early in the program. If the program metrics (e.g. Earned Value) begin to fail, both the customer and management teams must take immediate action to stop on-going development, and allow or force the engineering teams into top-down analysis and solutions. If they don't stop development, almost always, the program will be cancelled without delivery. Analysis on a large number of failed programs showed that if the management metrics show a downward trend for 6 months, the program will more than likely fail to meet cost and schedule objectives, and in a large number of cases fail to deliver.

The hardest decision the customer, management team or chief engineer will have to make is the decision to stop work. This is caused by the political trouble which follows. To build a system (large or small), the personnel in decision making roles must have the courage to stop work at the earliest signs of trouble, and enforce proper solutions be developed. This may lead to increased cost and schedule issues, but if the technical merit is properly achieved, the system will be deliverable.

A technical mess will never make it into operations and survive.

10.2.10 *Pitfalls Summary*

Whether or not it's a new system, whether or not reuse is in play, and/or an enhancement of an older system, it cannot be stressed enough, that bottoms-up engineer should not be used for systems development. Almost all development managed in this fashion resulted in a failure of one or more of the disciplines needed in system development.

The MDSE engineer should work top to bottom, whether it is an enterprise solution, system of system solution, or system solution, to a component in a system solution. Starting with the proper documentation, developing the proper requirements, architecture, OVs and SVs will result in the ability to develop new, or modify existing reuse solutions to provide a viable product to the customer [129].

10.3 Case Study: Classical Disasters in Systems Engineering

As we outline in earlier chapters, most new systems/solutions are extensions to older system/solutions. A large number of new systems are developed by the modifications of existing systems and software, were the software is sometime recoded, or adapted to run on new COTS hardware, software with incorporate of new commercial applications. A major issue affecting new systems is the need to enhance the older solutions with modernized COTS, new security apparatus and alike. A major pitfall affecting the customers, and the developers, is the need to "build more for less cost". Almost always, the solutions offered to the customer are refactored, or reuse of existing solutions using the current marketplace advancements in COTS hardware, software and new applications.

10.3.1 *The Classical Pitfall*

- The customer's expectation, when spending $1B, is that the new system **will be new**.
- The developer is trying to add more capabilities to their existing solutions sets, **without major changes** in their existing product lines.

10.3.2 *Typical Outcomes*

- Massive overruns, and/or
- Cancelled program.

10.3.3 Background Material

10.3.3.1 Axiom

Technical excellence will most always be supported with cost and schedule, but the inverse is not true. Working to cost and schedule, while failing to deliver on technical merit, will always lead to failure.

If you take care of technical merit, cost and schedule will take care of itself

This case study will focus on a program that was started, went to stop work, restarted, and then finished with a cancellation and no deliverables. This study will expose issues with both the customer and the developer.

10.3.4 Historical Reference

The customer operates multiple data centers geographically distributed, where for ten (10) plus years, the customer had initiated studies using multiple vendors to find a solution to an integrated system operating across geographical dispersed operational centers. In numerous cases, medium sized development programs were started and stopped due to the mergers in both customer and development organizations. Finally a formal competition was undertaken to establish a complete strategy to buy, integrate, and field a system of system solutions operating geographically 24×7.

The first phase, Concept Definition Phase (CDP), was initiated with four (4) vendors with extensive experience in the market place as well as building and fielding SoS solutions. The competition was executed with the four teams, where the outcomes were eventually shared and scored by the customer and vendors.

The second phase, Conceptual Prototype Phase (CPP), requested the vendors bid and build solution for a set of key capabilities identified by the customer and joint vendor teams. The prototypes were down selected and awarded to three of the four competitors. There was overlap in awards, such that first vendor won five of the prototypes, the second vendor received four, and the third vendor received two of the prototype capabilities.

While the prototype phase was underway, and with the third vendors realizing they were in a losing position, the third vendor developed a strategy to eliminate the other two vendors from the competition. In so doing, the third vendor established a "strategic integration center", and developed a system solution using mostly new COTS products from the marketplace, conducted a demonstration, and provided an estimated build cost for the SoS solution (for this review we label it as "The Marketed Price"). The value was so low that it was identified as "getting pregnant money" by one of the other vendor's Chief Engineers.

While the other vendors executed the CPP prototype developments, the contractual offering was awarded to the third competitor as an unsolicited award. To avoid lawsuits and challenges to the award, the customer invited all three CPP vendors to join together, and build the new SoS under the leadership of the third vendor. The forced marriage had numerous problems to overcome, namely, all three vendors could build the SoS, or portions of the SoS on their own. This led to an extremely intense in-fight to determine who was responsible for what pieces of technical work share.

The program was carved up into three major pieces: Program Management, System Development, and System Transition. The system development was subdivided into three main categories, with forced team integration on two of the three integrated development teams.

10.3.4.1 Contract Startup

The integrated development team provided a pre-contract SoS overview to the customer and the leads from the geographically dispersed operational centers. As part of the post assessment review, two of the four major pieces of the SE materials were found to be defective: Concept of Operations (CONOPS) and System Requirements. The Architecture and System Transition Overviews were graded by the integrated customer team as acceptable [78].

10.3.4.2 System Requirements Review (SRR)

With outstanding concerns from the SoS overview conference, the program started the staffing and development cycles. Rather than fixing the CONOPS and requirements, System Engineering decided to correct the concerns during the engineering cycle leading to Systems Requirements Review (SRR). To correct the CONOPS, SE decided to perform use case modeling with requirement derivation, rather than requirements decomposition and architectural assessment.

As system engineering was on going, the program management team developed the final cost of the system, pulling key personnel out of system engineering to help work the bid models, build plan, schedules and like artifacts to manage the cost and schedule of the program. The initial submission to the customer was a 100 % increase in cost to "The Marketed Price". This new cost, led to multiple cycles of interruption to the SE efforts, prior to finalizing the "Build to Price" (about 40 % higher than "The Marketed Price".

Leading into SRR, the SE efforts to correct the CONOPS and requirements failed to be achieved. Also, by SRR, the software and hardware developers were staffed and prototyping solutions for "high risk items". During SRR, the big four (CONOPS, Requirements, Architecture, Transition) were reviewed with the integrated customer team again, and again, the CONOPS and Requirements failed to pass approval.

This led to multiple meetings, and program realignment of key personnel, to provide more focus on CONOPS and Requirement corrective actions; however, both the customer and vendors didn't want to stop work for 3–6 months to correct the major discrepancies. It was decided to fix the issues during development activities leading to PDR, as was the decision leading into SRR.

10.3.4.3 Path to Preliminary Design Review (PDR)

Over the next 6 months, multiple cycles on use cases and requirements were performed in parallel with software and hardware development activities. The requirements associated with software and hardware activities were not fully vetted and approved for development. This led to bottoms-up development, where the S/W and H/W developers fell back on prior system development efforts and "built what they knew needed to be built".

The development schedule was revised to support critical path analysis, and a build plan to satisfy the customer's development objectives. Also, while the program management team and the Office of Chief Engineer (OCE) were building the new critical path schedule and build plan, the development manager formed a parallel team of engineers to decide what they would build in each iterative development cycles independently of the program management team and customer objectives.

At the 1 year mark, and 3 months prior to the PDR, the costs and schedule overruns were becoming apparent, as well as the gaps in the system's requirements and use cases. The new use cases were not helping to solve the CONOPS issues, and the actual development of software and hardware were radically off course relative to the critical path schedule and build plan.

The customer requested a corrective action plan, and program cost and schedule assessment be performed. The assessment was for both a cost and schedule, and a technical engineering assessment.

With the technical problems well understood by the OCE, the corrective action plan, for the technical merits of the system, was to stop development activities, and pursue a proper top down requirements decomposition analysis, and unfortunately, a scrapping of some of the developed software. The cost and schedule assessment showed a potential grow of cost of about 40 %, caused by rework and schedule growth.

With the pending dismissal of the development manager for building invaluable software, program management overruled the OCE corrective action for a 3 months period of time to allow the development teams a chance at integrating what was built to date. The hardware and software developed bottoms-up failed to integrate, and the technical fixes to the baseline were not achieved due to personnel shortages caused by support to the integration activities.

10.3.4.4 First Stop Work Issued

Fully 1.5 years into the development the program came to a stop. Sixty five of the staff of 325 were retained to evaluate and correct the technical merit issues, as well as the cost and schedule profile for the program. The OCE was scaled down to 4 staff members, a prototype staff of 15 members, with the rest of the staff working re-planning of both cost and schedule. During the first 3 months, the CONOPS was redeveloped by knowledgeable engineers and operators from the operational nodes, and formally approved by the customer. Over the next 4 months, the OCE and prototype staff correct the entire requirements set, and build of a fully functional prototype for the customer's end system vision, which was also formally approved by the customer.

In parallel with the requirements efforts, the prototype software development team working on the prototype, removed code developed by the original program staff, corrected software design deficiencies, instituted night builds using automated PPS&C checks, automated unit test runs, and integrated system startup tests. Their efforts took a non-functional H/W and S/W baseline, removed 30 % of the duplicate bottoms-up code, and delivered a prototype fully capable of operating in an operational node. With 2 months of coordination, the program was restarted 9 months after the initial stop work order [130].

10.3.4.5 Path to Preliminary Design Review (PDR) Redo #2

With approval from the customer for all of the big four (CONOPS, Requirements, Architecture, Transition), the program started re-staffing, with a requirement to be ready for a full PDR 6 months after restart. However, as with all developments there is a price to restart; namely, the customer was asked to include some new capabilities by the operational nodes since they were now 2 years behind schedule.

Unfortunately, the vendor work force, put back together in the forced marriage, came with issues. The prime was no longer in charge of the OCE, two of the sub-contractors were given new work share allowances smaller than the original pre-stop work contracts, but with more responsibilities to development. This led to some interesting behind the scenes negotiations between the program management and the vendors, as well as between program management and the new system engineering team, where both sets of negotiations were carried out without OCE oversight. By the time PDR was to occur, the customer observed the following major issues with the restart effort:

1. The new capabilities and resultant requirements development effort didn't match the approved CONOPS
2. The SE readiness was SRR quality, not PDR quality
3. The staffing and vendors "storming and norming" phase were resulting in inability to control engineering and development efforts
4. And most importantly, the customer's objectives were at risk again

Exactly 6 months after restart, and with PDR due, the program went back to stop work for the second time.

10.3.4.6 Second Stop Work Issued

Now, having started and stopped twice, the customer started accepting technical help for political reasons, and the vendor team was augmented by the addition of a fourth vendor, all with equal work share. Multiple technical meetings were conducted, technical requirements for development were reordered, and a new program plan was developed. This new restart was not going to solve the total objectives of the customer, but was going to integrate a new display tier and scheduling solution, and then integrating into the existing operational baselines.

10.3.4.7 Restart to Termination

Over the next 6 years, with multiple technical directors from different customer organizations helping, and four vendors with different agenda(s), the program was finally terminated with a cost overrun of 500 % of initial "The Marketed Price". The developed software and hardware solutions were scrapped, and the program terminated.

10.3.4.8 Case Study Assessment

There are multiple failures in this case study. They affect System Acquisition, Teaming and most importantly, multiple drastic failures in System Engineering. While there is a well-founded fact that "The customer is always right, because they are the customer", the truth lies in the ability to convince the customer with facts and data that the path to success should be with sound multidisciplinary systems engineering.

10.4 Discussion

MDSE is not new, although the terminology is just becoming "main stream." The Chief MDSE must ensure that his team understands the MDSE approach and all are on board with it. As in Agile Development, team dynamics are everything [116]. It will take much training at Universities and within organizations to grow effective MDSEs. Not all Systems Engineers are equal; you cannot just open up a can and make up an efficient, effective, well-tuned team that can bring a complex System of Systems program to market the meets requirements, cost, and schedule. It's hard

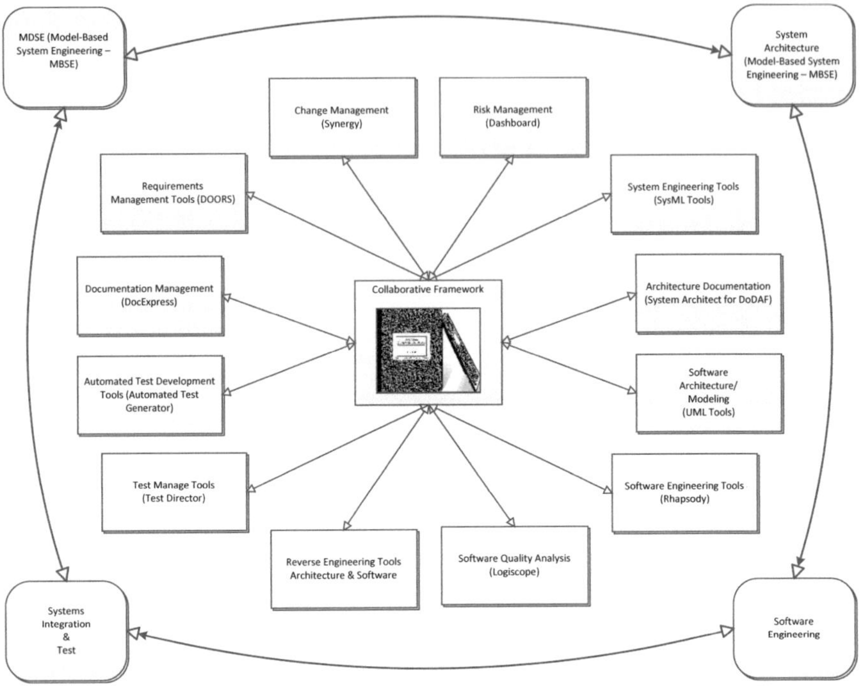

Fig. 10.3 MDSE collaboration framework

enough to accomplish this with the right team. With the wrong team it's nearly impossible. Figure 10.3 illustrates the MDSE collaborative process/framework. Figure 10.3 lays out the major disciplines that must converge Systems Analysis, Systems Architecture, Systems Integration and Software Engineering to provide an effective SoS design. Secondary (as shown in Fig. 10.3) all the other disciplines that must converge into MDSE.

Chapter 11
Conclusions and Discussion

11.1 Useful Systems Engineering

In the end what we want, what we need, is MDSE that provides value and is useful to the overall life cycle of System of Systems. Systems Engineering is important because it allows systems to be architected, designed, implemented, tested, deployed, operated and maintained with separable components at all levels of the system (element, subsystems, services, CIs, and components), such that the System of Systems can be managed, operated, and maintained at low cost over a long period of time. Some programs would not benefit from this approach, but those are few and generally small projects. As an MDSE, hopefully the following phrase is never used:

> There's never enough time to do it right, but there's always enough time to over.

This classic phrase described many, many systems, as engineers and managers fail to understand the importance and need to do the Systems Engineering right the first time. What follows is a discussion of what happens when MDSE is done right, and what happens when it is down wrong.

11.2 When Systems Engineering Goes Right

Almost all programs will encounter problems in development; however, when the Systems Engineering discipline is properly followed, and the full scope of the objectives and requirements provided by the customer are address across all disciplines of Systems Engineering, the programs are placed into operations. Most well engineered programs meet the cost and schedule targets, in addition to the technical capabilities desired.

Properly executed MDSE may encounter push back by the customer or company management, where the "do more with less" demands are encountered. The purpose of MDSE is to provide the necessary facts and data for the overall scope of the

J.A. Crowder et al., *Multidisciplinary Systems Engineering*,
DOI 10.1007/978-3-319-22398-8_11

program, and all changes being requested in the development. The negotiations between the MDSE and the customer/management teams typically flow smoothly when proper facts and data are presented.

Well executed programs establish and enforce the engineering processes and practices to be used during development. All new personnel added to a MDSE effort should be provided training on the contractual objectives, concepts of operations, requirement and architectural flow down, and procedures being employed on the program. While most organizations have standardized processes and practices, they allow the processes and practices to be tailored based on the program. All startup staff, new supervisors and new engineers being added to a development, should be training on the program specific process and practices. Well run programs do not allow engineers to start on the program, and continue operating as if they came from another program.

Well executed programs establish milestones with well-defined requirements and objectives for those milestones, as well as on-ramp and off-ramp plans caused by changes in direction, objectives, requirements, and like procurement changes. The entire purpose of following MDSE with top-down engineering discipline is to protect both the customer and development teams from drifting off course, and ending up with a failed program.

Some programs are pilot programs that will lead to the overall system of systems development and or integration. The MDSE should ensure that the scope of the efforts match the objectives of the procurement, and not allow unplanned growth to occur during development. This requires the MDSE ensure that the flow down of requirements are properly being implemented in the software and hardware baselines appropriate to the procurement. Incorporation of "nice to haves", or "we know this will be needed" solutions are not executed in a well-managed MDSE environment.

Programs properly executed lead to satisfied customers, company management, development teams, and a low stress work environment. It's the satisfied customer that, more often than not, provides you with more of their business, and referrals to other customers. It is for this reason the engineer should develop multidisciplinary practices, and incorporate them into their daily duties.

11.3 When Systems Engineering Goes Wrong

As has been discussed throughout this book, when engineers do not following MDSE practices, with top-down analysis and engineered solutions, the development organization becomes exposed to higher risks, and potential for failure.

When program staff is not properly trained on the customer objectives, requirements, program plans, processes and practices, the system engineering effort will take on a flavor of "multiple different programs within a program". The development will enter into a reactionary engineering culture, where development is fragmented across different standards and solution sets, all leading to extensive risk, cost and schedule growth, and potential program cancellation.

The decomposition of the system should be uniform across the specific development, which will depend on the depth of the procurement. A simple prototype may have a few tiers of decomposition; where as an Enterprise of System of Systems may have dozens of tiers. Skipping tiers of decomposition typically leads to failure, but is acceptable when managed properly on smaller programs. An improper decomposition will eventually lead to bottoms-up and/or reactionary engineering during integration and test phases, where massive cost and schedule overruns will impact the fielding of the system.

The practice of reuse product lines comes with risk in new developments. More often than naught, the customer may have selected an organization to perform a new development to add diversity to his vendor base. While the selected vendors may have existing reuse product lines, the MDSE should ensure properly requirements and architectural flow down occurs. This allows the development staffs to select well documented solutions from the reuse libraries.

Chapter 10 outlined bottoms-up development, where reuse should come with libraries of linked artifacts that enable the selection of reuse solutions for inclusion in a new development. Taking for granted, that the older solution solves the new procurement with minor changes, is a frequent cause of program failures. The use of a product line, with changes to support a new development, typically leads to bottoms-up engineering, which then leads to reactionary engineer and then reactionary management and eventual program cancellation. In numerous cases:

- The reuse product lines are not up to the current security posture required of a new system
- The reuse product lines' COTS products have become obsolete in the market place
- The older human-machine interfaces are replaced by EEOC and/or disability act changes
- The system will not be maintained by the developer due to customer procurement constraints, etc. It is for these reasons that many developments have extensive cost overruns and/or failures when using reuse product lines.

Throughout this book, examples of failed developments were used to outline why System Engineering should employ multidisciplinary solutions to the daily activities of the development staff. A failure on a prototype garners low attention, but a failure on a System of Systems development becomes a marketing nightmare for the organization and in some cases the entire company.

11.4 A Look at Agile Systems Engineering

As we move toward more complex SoS designs and as Agile Software development becomes even more prevalent, Systems Engineering must adopt new philosophies and paradigms to keep up with the evolutionary nature of modern systems design. The MDSE must embrace this change and develop methodologies and structures which allow rapid and accurate evolution of the Systems Architecture as the SoS program progresses through its life cycle [131]. The evolution to Agile Systems Engineering is a natural progression from the pre-industrial age, through the industrial age, and now into the information age. The information age can also be thought of as the "Knowledge Age," with the understanding that data by itself means nothing. Data must be turned into information (data with context) and then into knowledge to support decision analysis in modern systems. The MDSE must be on the leading edge of this "Knowledge Paradigm."

11.4.1 The Knowledge Paradigm

Knowledge is an orderly synthesis of information; it is an abstraction of our understanding gained through experience and study. Information is an orderly collection of facts that form databases (whether in a computer or our heads); it provides input to our processes. Information Technology is the means used to gather and manipulate data to form knowledge [132]. Wisdom depends on patient osmosis. Before the industrial age, communication was transferred completely through written and oral means. Improved communications improved communications and became that catalyst for change throughout the world. This is illustrated in Fig. 11.1 in a hierarchical process-product system.

During the industrial revolution, communication and information generation was enhanced through the use of steam power. Steam powered ships and locomotives allowed newspapers to be delivered much faster and to areas of the world not previously accessible. In 1814 steam powered ships and locomotives arrived with the first copy of the Times in London.

After the industrial age, technologies rapidly increased, accelerating information to the present where we are becoming overwhelmed with data/information (i.e., Big Data concerns). It has become all too easy to copy data and forward it to multiple people. The questions then become, do all those receiving data/information need it; was it wasted? Waste identifies low quality in that wasted information reduces the amount of time available to produce useful information. In addition, low quality

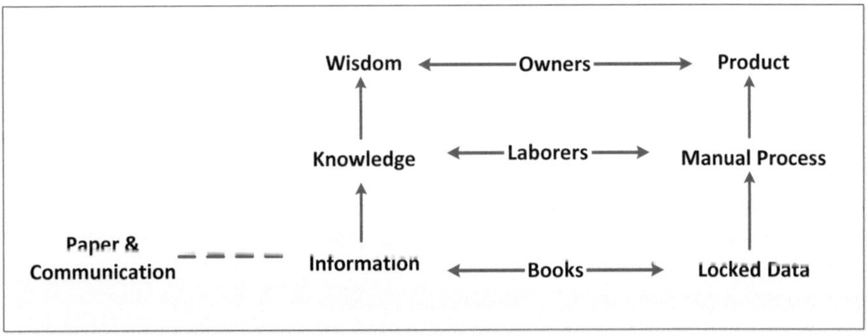

Fig. 11.1 Pre-industrial business communication and knowledge paradigm

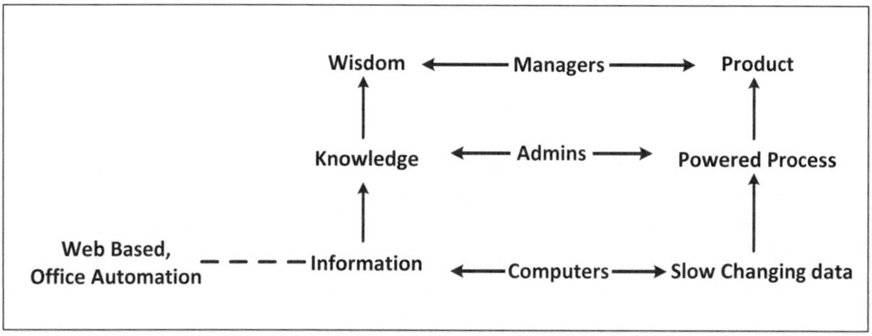

Fig. 11.2 The post-industrial business information paradigm

requires rework and therefore, lost opportunities to improve due to the time invested to rework. However, the proliferation of information has enabled the development of expertise. It is possible to find data on almost any subject across the Web. Figure 11.2 illustrates the change as computers, information systems, and the Word-Wide-Web have displaced books as the basis for information; administrative assistants have displaced laborers as the predominant component in society; and managers have displaced company owners as owners of the product and holding the "power."

11.4.2 The New Paradigm: Horizontal Integration of Knowledge

Perhaps the simplest example of how power is being shifted by computer technology is to consider communication rights. The telephone (including cell phones) and email are indispensable components in getting work done. This becomes painfully evident in third world countries where telephones, cell towers, and internet connectivity are scarce; there are no beepers, no voice mail, and no email [126]. In modern companies workers have communications rights, including long distance and international communication through email and web-based connectivity without asking permission. The engineer can now meet colleagues electronically and develop relationships without ever meeting them fact to face (although they can meet virtually face to face with video conferencing).

This newfound freedom in communications has provided a plethora of opportunities. Web sites can be found that are full of any information you might be looking for. News events are published to the Web as they are happening. Streaming video is available across the world on any event. Suddenly we are living in an electronic village. The shy person is not afraid to speak up, as they are secure in their electronic village. Also, distance no longer prevents casual conversation worldwide as you can simply open up your virtual cottage window and simply chat with the electronic passerby. In addition, one can browse electronic databases from anywhere in the world like walking a trail through the local park.

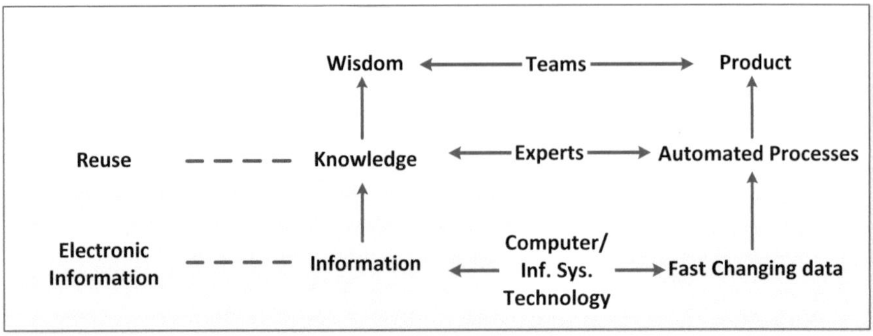

Fig. 11.3 Business paradigm for the knowledge age

Current computer technology enables us to redeem the time we need to discover what's out there. It provides a broad horizontal integration of the workplace and the rest of life. Within an organization, the horizontal integration is more focused. Just as the manager replaced the owner at the turn of the century, today the team is replacing the manager. Teams can be formed on the fly as engineers and scientists find one another over the network, identify their common vision, and then they can disband once they reach their objective. The bureaucracy, that integrated the clerks, is being replaced by the electronic network of experts [133] and systems engineers. The middle manager will have a new role on the team: facilitating the brain storming and design sessions (possibly driven by QFD analysis) and keeping relative focus for the team. This is illustrated in Fig. 11.3, where teams replace the manager and experts replace the clerk.

However, there is another important distinction between Figs. 11.2 and 11.3. The Industrial Revolution enabled the rapid proliferation of information, and now the Knowledge Revolution enables the rapid reuse of information and knowledge. There is need for what are known as Functionbases [56], which captures rules for using information and knowledge. In this manner a high level of organization can be imposed upon the information for rapid reuse that provides history, background, and context for the information. Automation then enables members of teams to use one another's functions with ease. Less time is wasted on the detail of how; more time is spent on the details of why [133]. This is how teams self-manage, replacing managers as owners of the product.

11.4.3 A Simple Functionbase Approach

The paper paradigm pervades every aspect of our lives. Efficiency improvements often lead to the generation of more paper as quality improvements are sought. The purpose of the Functionbase is to capture the process using information technology, eliminating paper capture of information and rendering the process repeatable and reusable. Using paper, and even using thousands of disparate files in "electronic collaboration" paradigms, causes us to be awash in information and dependent of quality checks on the product upon its completion. Even the use of standard office automation products provides little help. Many design documents and information is lost in a sea of electronic folders from individual engineers in an ocean of directories. Often information is lost, as no one knows where the information was stored, under what directory and what naming convention. This process is little better than paper reports in separate filing cabinets.

A simple example is provided by the spreadsheet. Early spreadsheets were databases that helped in the layout and basic organization of information. One early innovation was to add Macros that could repeat specific user operations. All that was required was training on the computer application the particular operations, and it could then be repeated on demand. Voila, automation. Ensuring the Macro was correct is putting quality into the Functionbase. The product inherited the quality and did not require checking. Checking had become a waste, quality had been improved.

The automation for MDSE engineering functions and disciplines is similar. For functions, such as Business/Mission Management, Command & Control, or Ship/Truck scheduling, build a database of functionality and append a Functionbase, such that the design is automated. What's more, the process tool must be self-documenting. If the MDSE has to "copy" information into a flat-file report, then there is opportunity for error and quality control checking will have to be reinstated. The team need only check the original Functionbase knowing the new data propagation will be correct. This method has been tested using MATLAB® and METADESIGN® for the design of a Metrics Analysis Systems. These two tools easily integrate together to form a Functionbase that captures the Metrics analysis process cradle to grave. The following attributes have been noted [134]:

(1) ***Productivity*** was increased up to tenfold.
(2) ***Repeatability*** enabled inexperienced engineers and team members (including the customer) to reliably reuse the design.
(3) ***Shareability*** has enabled the Functionbase to be the contract deliverable.
(4) ***Verification and Validation*** engineers could now review the Functionbase for design methodologies, codes, results, etc.; a complete self-contained package.

11.4.4 Engineering Tools: The MDSE Sandbox

Engineering tools like MATLAB® and IDL® serve two important functions. They can be used as executable specification to software engineers and provides the design engineer with a sandbox for experiments as illustrated in Fig. 11.4. The use of engineering tools as executable specifications has many advantages. Code and results produced from such tools can be used to validate Unified Modeling Language

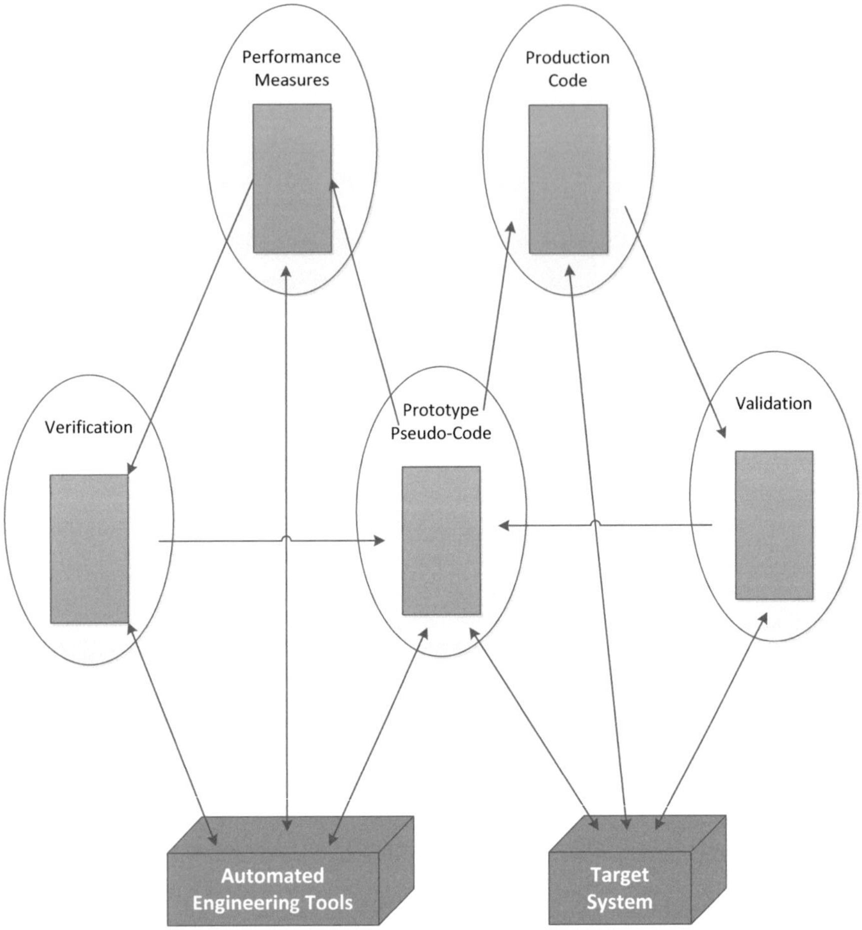

Fig. 11.4 MDSE automated engineering tools for integrated analysis and design

(UML) flow diagrams, utilizing the engineering tool code as pseudo-code to validate the UML design documentation for the subsystem. The advantages are:

(1) *Single Representation* of the real-time code requirements. The Functionbase from the engineering tools becomes pseudo-code to the target system software engineers and an executable analysis code to the systems design engineer. There is no redundant data to confuse the team when it comes to resolving design ambiguities.
(2) *Mathematical Notation* makes it easy to correlate the design code with the mathematical derivations. This facilitates verification of requirements.
(3) *Embedded Prose* works hand-in-hand with the natural readability of the engineering tool's mathematical notation. Prose is easily embedded into the code to provide further explanation as to what the code means.

(4) *Executable Code* is the icing on the cake. Because the code from the engineering tools provides performance data, analyzing the performance implicitly ensures the quality of the code as a pseudo-code for the target system software engineers.
(5) *Independent Review* of the engineering tool code is a natural attribute of the method, afforded by the target system's software engineers. It ensures a methodical, independent review of the code design as it is re-hosted onto the target system.

Finding tools that the team all agrees on can be one of the hardest tasks. Many tools provide a means of integrating analysis code and real-time pseudo-code. They provide the teams with a sandbox by virtue of their rich function libraries. Whether MATLAB®, IDL®, or Satellite Tool Kit®, or other tools, they can provide the engineer with an inexhaustible supply of ideas and experiments, especially when user groups are taken into consideration.

Designs that once required specialized code on servers can now be produced using existing library functions on the desktop. This has collapsed analysis time by up to 90 % in some instances; this productivity improvement is what makes product improvement feasible. Without it, change becomes a major undertaking for even the smallest improvement.

11.4.5 Automated Process Documentation

METADESIGN® or other object-oriented design tools like System Architect®, or Rapsody®, allows creation of Functionbases that manage knowledge about the design throughout the systems lifecycle. However, the components of the Functionbase are best understood by reviewing how it should be used in conjunction with the other tools. Recurring Design is the most common use. This occurs when a new configuration requires a change (say to a Mission Management system). Figure 11.5 illustrates the process utilizing the Functionbases. Starting at the top of the *Recurring Design* tree, this first page that appears would be instructions on what scripts to run in what order for the engineering tools associated with the Mission Management models. This is accomplished with hyperlinks on the first page to other pages within the Functionbase. For example, the first link should be to the *Configuration Data* page. Entered onto this page is all of the data required to complete the design. The next link should be to the engineering tool script created for the design model and etc., until the design is complete [135]. During the design sequence for the changes, the engineering tools will generate text containing the new set of data for inclusion into the Functionbase for the new design (the changes being made). When run, the new scripts that generate the *Performance Measure Data* will use the *Data Dictionary Tree* and the *Pseudo-code Simulation*. In this manner the files used by the target system software engineers are automatically checked through the performance analysis.

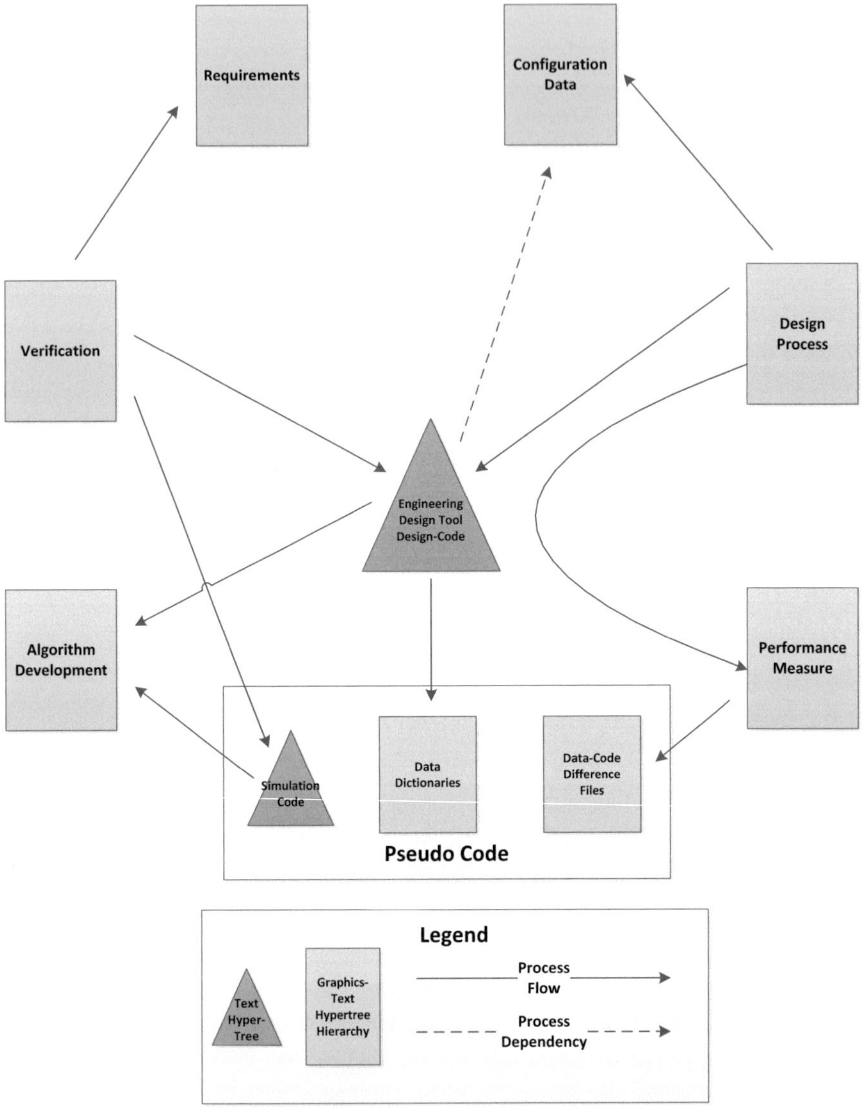

Fig. 11.5 Generalized functionbase reuse

On completing the *Recurring Design* exercise, the target system software engineers can access the pseudo-code trees and export the necessary data files. Included are *Difference Files* that show exactly what pieces of data have changed. This includes a change log with references back to the design discrepancies that invoked the change. The log is important in that it provides a historic trail for new members to understand the design and coding evolution. Without such a trail there is no corporate memory and old mistakes will be repeated.

11.4.5.1 Verification Matrices

Verification Matrices are included to enable any user to trace how the requirements are being met. In the QFD sense there is also a link to the *Performance Measures* to ascertain how well the requirements are being met. Links are again placed on the page so the user can "jump" from a requirement to the particular page within the Functionbase that addresses that requirement.

11.4.5.2 Algorithm Development

The *Algorithm Development* tree provides additional information to enable the user to understand the design and pseudo-codes/target system software. Again links are provided for use to jump through the Functionbase. This section might also provide the scripts that run the pseudo-code and the notes that help explain certain design decisions (i.e., trade studies and sensitivity analyses). Again, without this errors will be repeated as team members relearn old lessons.

11.4.6 The MDSE Engineering Design Reuse Problem Solution

No one wants to waste time reinventing the wheel, and nobody wants to admit that errors are repeated on a daily basis. Such a state of affairs is wasteful and lacks quality. The MDSE's desire is to have a library of code to be reused with all available design and test information available easily. But before jumping on the reuse bandwagon, consider the classes of reuse.

11.4.6.1 Software Libraries

Software Libraries contain annotated code for reuse (e.g., freeware). This type of code can be helpful when it works first time and when proof of correctness is not a requirement. But for most software in most companies, software libraries might be better named "junkyards" as functionality and correctness cannot be guaranteed.

11.4.6.1.1 Designed for Reuse Functionbases

Designed for Reuse Functionbases contain more than just code. Included are the knowledge of the behavior, interfaces, design models and proofs. This constitutes the data required to certify the code for use.

11.4.6.2 Domain-Specific Solutions

Doman Specific Solutions are complete hardware-software solutions. The home PC delivered with preloaded operating system and office automation software is an example.

11.4.6.3 Knowledge Management

Discussed here will be the second class: Designed for Reuse Functionbases. For this class, the reuse requirements must be carefully defined.

First, knowledge must be defined and included in terms of a mathematical basis. This is why the reuse junkyard is so hard to deal with, as the mathematical basis is not known; therefore, its behavior within a system is unknown. The need for a mathematical basis is the reason engineering tools, such as MATLAB, as an executable functional specification are useful. The engineering tool pseudo-code syntax is mathematically based, and provides insight into the mathematically basis for the design.

Second, interfaces have to be completely defined to avoid errors that involve data flow, priorities, and timing errors at both the highest and lowest levels of the system [104]. Common experience shows interface problems to take over 75–90 % of the errors found after implantation of reuse code with conventional techniques. This drives up cost due to all the investments in manufacturing Interface Control Documents (ICD's). Robust error checking is an essential design-for-reuse feature, as it precludes this type of expensive error. Code correctness must be assured before compiling.

Third, design proofs are required to preclude system errors. A prior verification of code precludes expensive system bugs, such as instabilities. The design proof leverages off the mathematical basis; algorithm correctness must also be assured before compilation.

Four basic tools are required to provide the attributes for a real design-for-reuse paradigm:

(1) A distributed hypermedia tool for workstations to allow creation and tracking of Functionbases.
(2) Engineering design tools like MATLAB, IDL, and Satellite Toolkit for building pseudo-code simulations and prototypes.
(3) Mathematical Equation tool, like Macsyma® for building and encoding mathematical models.
(4) A Case tool used to implement the robust code building routines, needed to preclude errors at their source.

The relationships between these tools are illustrated in Fig. 11.6. The reuse methodology integrates the behavior of the two ends of the system. The engineering tools form a mathematical basis for capturing the system's algorithmic behavior. The case tools have a mathematical basis for capturing the target system's behavior. Experiments can quickly be run in the sandbox before time is spent completing the design. The Data Dictionary can be built that complements the engineering tool's functions, and the target code re-ingested into the engineering tool (in actual code C++, JAVA, etc.) for verification. Tests can then be devised in the sandbox for running on the target system for final verification and validation.

11.4.6.4 Provably Correct Code

The engineering tool's mathematical basis makes it an ideal candidate for experimental design. Part of the reason for this is that there are generally no typing constraints. Conversely, code for target systems require strong typing, hence the Data Dictionary (see Fig. 11.5). The experimental nature of the engineering tools is also used at the other end of the lifecycle in analyzing data from the target system (see Fig. 11.6). In between these two lifecycle ends is the Case Tool to bridge the gap between the functional architecture and the resource architecture of the target system.

The functional architecture is design by function hierarchies (called FMaps) used by the engineering tools and type hierarchies (called TMaps) implied by the engineering tools and used by the Case Tools. Three primitive control structures are used. There is one for defining dependent relationships, one for defining independent relationships, and one for defining decision-making relationships. A formal set of rules associated with these are used to remove design errors from the maps. Because the primitive structures are reliable and because the building mechanisms have formal proofs of correctness, the final system is reliable. Furthermore, all modal viewpoints can be obtained from the FMaps and TMaps (e.g., data flows, control flows, state transitions, etc.) to aid the designer in visualizing the design.

The engineering tools reuse value is in its mathematical basis. The reuse value of the Case tools is founded in its separation of functional and resource architectures. Once the functional architecture is defined, it can be used on any target system, whose resource architectures are fully reusable by the engineer, when they match the

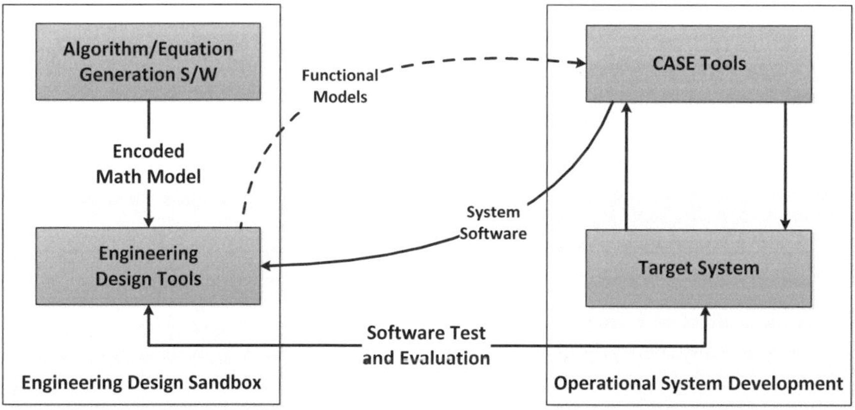

Fig. 11.6 MDSE design for reuse methodology

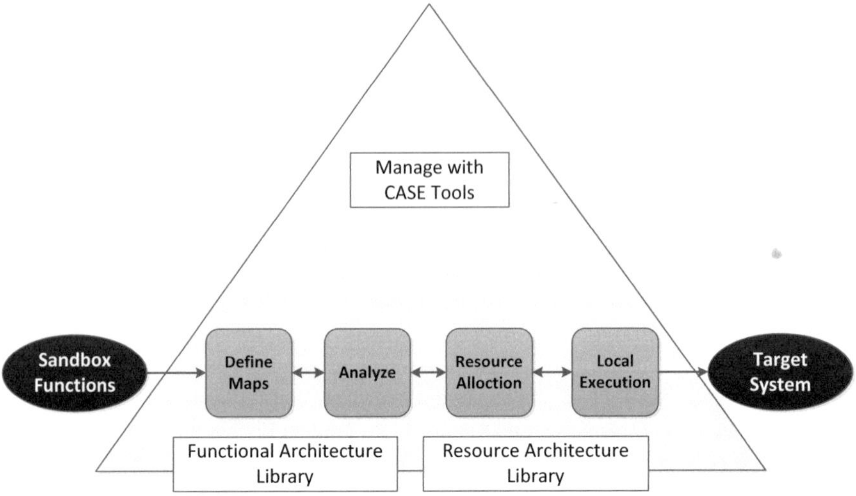

Fig. 11.7 Bridging the sandbox and the target system

engineering tools functions. Put another way, reuse requires two Functionbases: the engineering tools functions and the case tools functional architectures that capture the system behavior. These two Functionbases capture all the information required to complete specificity of the system, no more, no less. All that is required is the case tool that integrates and analyzes these architectures. This is illustrated in Fig. 11.7.

The analyzer is used during the definition of the TMaps and FMaps to test for consistency and logical correctness before placing the maps in the library. Templates of the particular target system are built and populate the resource architecture library. The Resource Allocation integrates templates from the library and then automatically generates the required source code. Run-time performance analysis of the code can be verified locally to ensure it meets the constraints of the target system (e.g., timing). A synergism is realized by using the engineering tools and the case tools together. They provide a design process with built in quality. When combined with the QFD methodologies, this paradigm provides a framework for automation, aiding the organization in defining and implementing process improvements [58].

11.4.7 Groupware for Knowledge Management: The MDSE Electronic Notebook

The Functionbase is structured to look like a technical memorandum at its top level. This enables anyone in the team to reuse the Functionbase as an electronic memo, and play what-if games to better understand the design. It also enables rapid-generation of the design, because the process is built into the Functionbases. Careful management ensures quality with the process (i.e., quality is built into the

Functionbase); therefore, time is not wasted checking the pseudo-code files or checking the performance analysis. The corollary for all this is the notion of a hypermedia, electronic engineering notebook. This MDSE Electronic Engineering Notebook (EEN) includes the software tools and codes, such that an engineering activity performed using the EEN can be repeated without additional knowledge. It contains the information on what it is, what it means, and how to use it. This Knowledge Management System provides the hypermedia capability to provide the functionality described above.

The EEN requires no coaching to use, and includes tutorials to aid understanding. The objective is to enable complete reuse of the Functionbase by a first time "functional stranger." The architecture required to support the multi-functional approach is illustrated in Fig. 11.8. The EEN paradigm has some important attributes:

Process Objects are directories containing all the files and directories required to complete a discrete design or analysis. The object is a collection of linked text, graphics, and applications.

Transparent Network Objects enable team members to browse one another's Process Objects. Virtual documents can be built and printed if required, using links that traverse the network to integrate Process Objects.

Process Automation is provided using a scripting language. The EEN can be taught the process that such that designs are automated.

Functionbase Security is provided to ensure data integrity is not violated.

Anyone who understands the above should see that this is infinitely doable utilizing today's Web Services and Java Scripting, coupled with Office Automation Tools that include hyper-linking capabilities.

11.4.7.1 Required MDSE and Knowledge Management Functions

To support the MDSE team, the EEN must contain a broad range of user functions. Two classifications can be made: engineering functions that support the program, and knowledge management functions that support the engineer. These are, of course, correlated, as illustrated in Table 11.1.

Some of the major categories for MDSE Required Engineering that must be captured by the EEN are described below:

11.4.7.1.1 Interactive Experiments (Sandbox and Analysis)

Interactive Experiments refers to the process whereby an engineer can create a "living" notebook in an on-line environment to retain analytical results and interleave comments and observations. This serves as a replacement of the classic Engineer's Notebook. This is accomplished in a real-time environment while analyses are actively being performed. This also allows creation of hypertext links to any other relevant material to provide a dynamically growing structure of cross-linked reference material. Analysis differs from the sandbox only in terms of rigor.

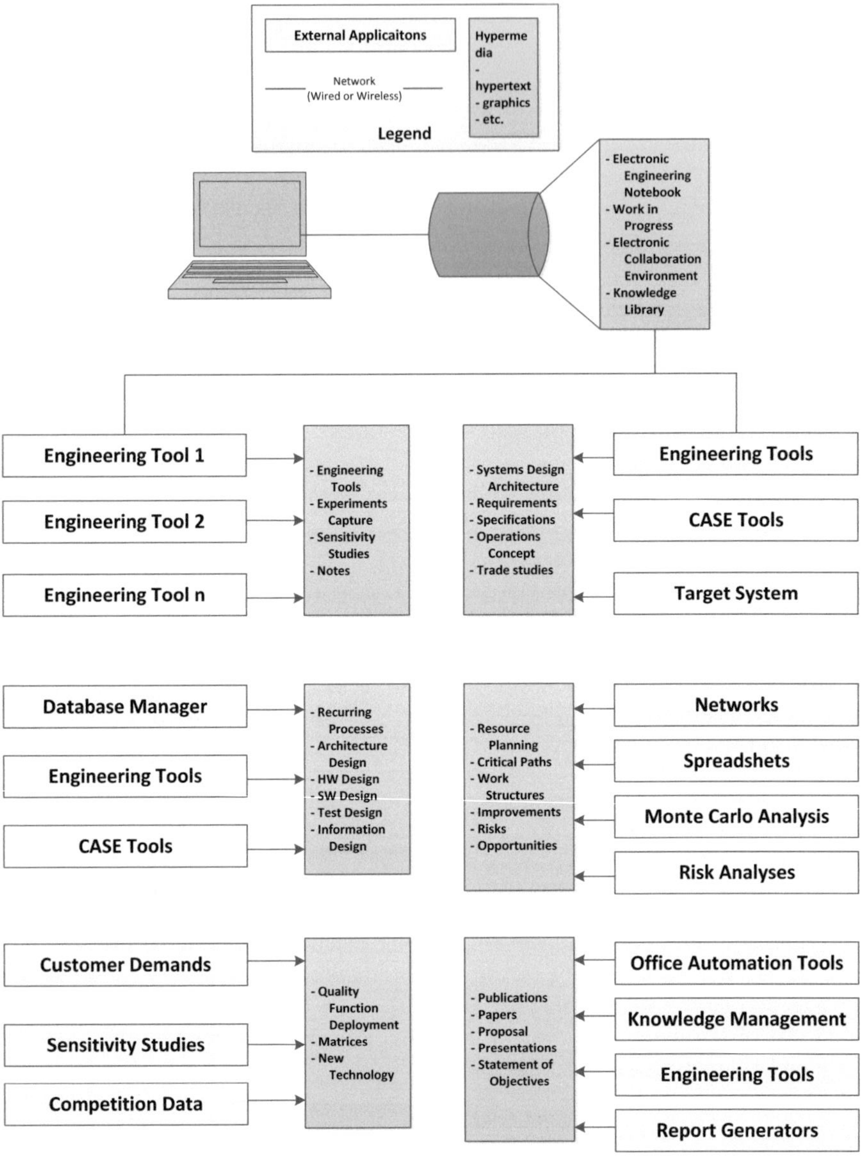

Fig. 11.8 Architecture for the MDSE electronic engineering notebook

11.4.7.1.2 Recurring Engineering (Automated Processes and System Data Configuration)

Recurring Engineering refers to engineering tasks such as performance analyses, design analyses and the measurement of performance for quality assurance, all of which can be automated.

Table 11.1 MDSE electronic engineering notebook functionality

	Knowledge management functions								
Required engineering									
Sandbox	X	X	X		X			X	X
Analysis	X	X	X	X	X			X	X
Automated processes	X	X	X	X		X	X	X	X
System data configuration	X		X			X	X	X	X
QFD	X		X	X	X		X	X	X
Resource management	X				X	X		X	X
S/W designs	X	X	X	X	X		X	X	X
Design release	X		X	X		X			
Presentations			X	X	X	X		X	X
Papers/memos/directives	X	X	X		X	X		X	X

11.4.7.1.3 Non-recurring Engineering

Non-Recurring Engineering refers to three domains. First the Functionbase, that contains the QFD matrices governing the integration of the organizational elements and the incorporation of new technologies. Second, resource planning requires worksheets, networks, and analyses essential to the determination of the work plans, the integration and management of resources, and the incorporation of improvements. Third, the design phase (S/W Design and Design Release) which are all those activities performed by the engineer, such as building a Functional Architecture Library.

11.4.7.1.4 Publications (Presentations and Papers/Memos/Directives)

Publications are the release of engineering requiring navigation paths to select subsets of information. This can be performed in the knowledge management domain (Web Services) or can involve Case Tools and may be published in electronic media form.

The major categories for Knowledge Management Functions, shown in Table 11.1 are:

11.4.7.1.5 Data Management

Data Management involves integration of both graphics and text into a hypermedia Functionbase. Data access must be capable of being automated.

11.4.7.1.6 Function Management

Function Management involves the development and structuring of procedures. An example is code management (Configuration Management), where code should be captured within a tree and executed directly (i.e., through a button on the screen).

11.4.7.1.7 Linked Data Structures

Linked Data Structures are methods of data organization, which consistently allows the monitoring and modification of data interrelationships and interdependencies. The structures are composed of trees that form objects and navigation paths across the trees.

11.4.7.1.8 Unified File System

Unified File System means a data item exists in one place only. There are no electronic copies (except for back-up), as this would compromise the Functionbase. All processes refer to the one copy.

11.4.7.1.9 Automatic Notification

Automatic Notification is the ability to automatically notify a designated user, or list of users, when a particular, selectable, event has occurred. This is used when a browser leaves notes and comments, or when data someone else is dependent on is changed by the originator.

11.4.7.1.10 Quality Assurance

Quality Assurance ensures engineering is not released until all Quality Characteristics have been verified. Because the QFD established the minimum performance requirements, all which are required, is the knowledge management system within the EEN to check that each Process Capability Index. Anything less would indicate nonconformance.

11.4.7.1.11 Tool Interface (Case Tools and Engineering Tools)

Tool Interface is the ability to invoke Case Tool or Engineering Tool programs by spawning a separate process. This includes the ability to interface with the program by sending input to and receiving input from that program, from within the EEN. The data so transmitted (and stored in the EEN) must be both alphanumeric and graphical. This program should be capable of being run in an interactive mode as well as batch.

11.4.7.2 Using Knowledge Management

The Need to Use Metaphors

As with any paradigm shift, Knowledge Management has been received with skepticism. The perception that the Knowledge Management paradigm is an improvement is more readily perceived when it is packaged within a familiar metaphor, hence the Electronic Engineer's Notebook name. Once engineers begin to experience it, and management sees it, the community will accept it as a productivity booster.

11.4.7.2.1 Knowledge Management as a Life Cycle Tool

Knowledge Management will gain acceptance across the industry, but not without some pressure being applied to the engineers that use it. This can be attributed to the general reluctance to change we exhibit as humans. Making the Knowledge Management paradigm broad in application will help in that an engineer will be able to use it for a task that is personally comfortable (e.g., making viewgraphs). Once on the learning curve, the engineer will grasp some of the other more subtle aspects of the paradigm with growing experience.

11.4.7.2.2 Advanced Communications

Being able to pass reusable knowledge to a peer will have advantages on both the Intranet and Internet. Whole trees are treated as objects, and can be shipped out to teammates in organizations in any geographically diverse location, or are used in conjunction with a Virtual Development Environment. The advantage this provides is that teams now become a virtual team. The teams will appear to operate "elbow to elbow" even though they seldom meet "face to face." The hypermedia aspect afforded by the Web and by Web Services makes this process doable today.

11.4.7.2.3 Emphasis on Learning

Given robust design algorithms, a hypermedia Functionbase described here will accelerate engineering such that more time can be spent finding ways to improve the design. Presently, many hours are spent rerunning codes, collecting data for presentations, and making reports, all of which could be automated. Time is redeemed for more erudite pursuits enabling the engineer to focus on improvements.

11.4.7.3 Knowledge Management and the MDSE EEN

Computer and Information System technology has provided us with a new Knowledge Paradigm based upon the power to extract data and functions, such that the functional stranger can use it. It has also enabled new horizons for communicating new ideas and processes such that teams are becoming virtual. Historically, the customer and the various teams have been geographically remote, requiring the periodic design reviews that dictate a sequential design mode. Now, given this ***Knowledge Management Paradigm***, the team can be extended and modified to include the customer, giving them insight into every aspect of design and implementation. Consensus, which was formerly developed post-priori using design reviews, can now be developed a priori using real-time communications of Functionbases.

11.5 Organizational Changes for MDSE

People dislike change. It takes them out of their comfort zone. Human nature demands continuity and consistency. For the MDSE, these factors must be taken into consideration in our engineering if continual improvement is to be realized. There are telltale signs that indicate when change is a natural part of the process [56]:

Quality is measured to prove the customer's issues are being met and to provide the basis for forcing change. Changes required for Continuous Product/Process Improvement (CPPI) to become an everyday affair.

Quality is implicit and is as much a part of the engineering as correct math; if the team cannot generate correct math, then the organization needs a math department. The same is true for quality. So if the organization has a quality department, **it indicates engineering processes that are not robust cannot integrate change**.

Quality is related to training and education because change is perpetual. Quality is a race to be run, not just and objective to complete. Therefore, it is likely that at least one fifth on an engineer's time will be spent in training and education. Without it, the Interactive Experiments required to find improvement opportunities will have low yields [136].

11.5.1 The MDSE Organization

New organizational structures are required to facilitate that allow the MDSE teams to own the products and resources. In some respects, this is a return to Frederic Taylor's Scientific Management where the team is the industrial engineer, determining how the resources can best be utilized for process-product improvement [137]. These resources may be allocated to education, faster hardware and software, additional team members, etc. This may require an organizational focus on long-term profit before commitment can be made, as the cultural change involves the engineer as well as the manager. Some short-term losses will no doubt be incurred as the team learns to take advantage of the new paradigm.

11.5.1.1 Robust Designs the First Time

The Expert Systems Designer is our concept generated by all the ideas discussed in this book and is graphically represented in Fig. 11.9.

The attributes of the Expert System Designer are described as follows:

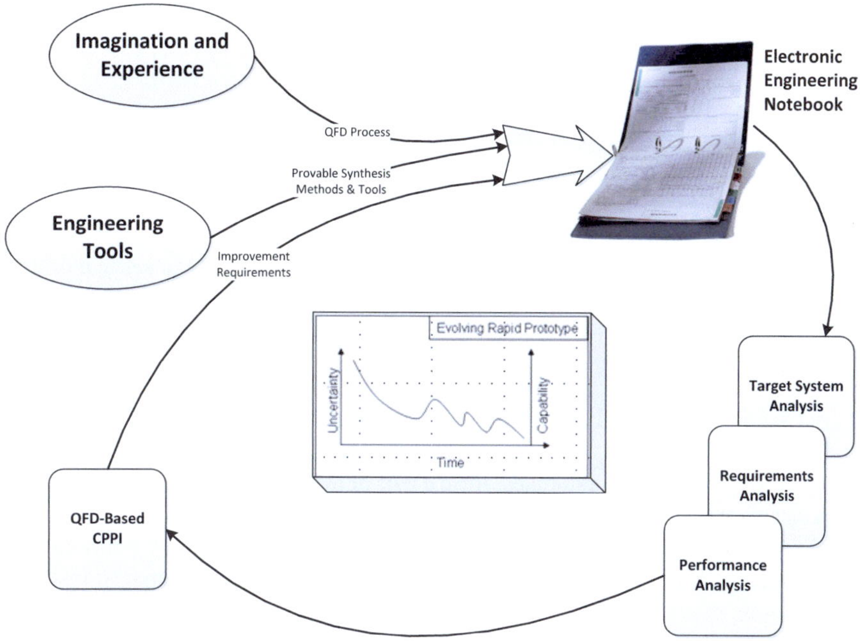

Fig. 11.9 The Expert Designer: Agile Systems Engineering

11.5.1.1.1 Quality Function Deployment

QFD provides a means of capturing an organization's existing knowledge for reuse by less experienced engineers. In this manner, less time is wasted generating the product that best meets the Customer Demands as old mistakes are not repeated and the wheel is not redesigned. This could even be taken to the point of using Functionbase interrogation to find the best QFD to fit a new set of Customer Demands, or even amalgamating components of several QFD Functionbases to forge an even better fit. All that would then remain is to check that the design is ***Right-First-Time*** through the performance estimation.

11.5.1.1.2 Provable Synthesis Methods and Tools

These provide the foundation for design without error. The fact that most errors are repeated is not new; this is why code libraries have been made for reuse. What is required is to integrate the knowledge about repeated errors to effect robust error checking, hence the Case Tools. Also, there is the question of the quality of the requirements. This requires the analyst to question his own thinking through experimentation. The Engineering Tools provides the Engineering Sandbox with rich function libraries. The two tools together provide a synergism where new ideas can be developed and implemented ***Right-First-Time***. The method for eliminating errors is aggressively preventative in nature, not corrective as post-priori bug elimination is wasteful and uncertain.

11.5.1.1.3 The Electronic Engineer's Notebook (EEN)

The EEN integrates the QFD, methods-tools, and improvement requirements. It enables the functional-stranger to rapidly repeat an analysis or regenerate a design. It is the Notebook that will enable the engineer to spend less time on clerical work and, through automation, more time on invention and design.

11.5.1.1.4 Analysis of Performance

Analysis of performance estimates, form a part of the EEN. Development of models provides the means to test and tighten performance estimates and validate what we think we know. This provides the basis for change; changing what we don't know is always dicey. With proven knowledge, change can be made in a managed fashion.

11.5.1.1.5 Continual Process-Product Improvement (CPPI)

QFD Based CPPI is the integration of new technology, concepts, and knowledge. Decision-making is based on Cp measurements and how they will be affected. Such a rational basis is a prerequisite to paradigm busting. QFD is a discipline that ensures the product does not suffer paradigm paralysis.

11.5.1.1.6 Evolutionary Rapid Prototype

The Evolutionary Rapid Prototype describes the capability realized by the Expert Designer concept. Evolution is a gradual process in which something changes into a different, and usually more complex, form. Rapid means moving and moving swiftly. Prototype is an original type, form, or instance that serves as a model on which later stages are based or judged. Therefore the Evolutionary Rapid Prototype means each process-product is a basis for the next. The process-product is ever changing and always improving.

11.5.2 MDSE Continuous Improvement Paradigm

Three of the Required Engineering Domains described in Fig. 11.9 were shown in Fig. 3.4 and replicated in Fig. 11.10 in order to illustrate their relationship to each other. The MDSE's time might well be split evenly between each domain to

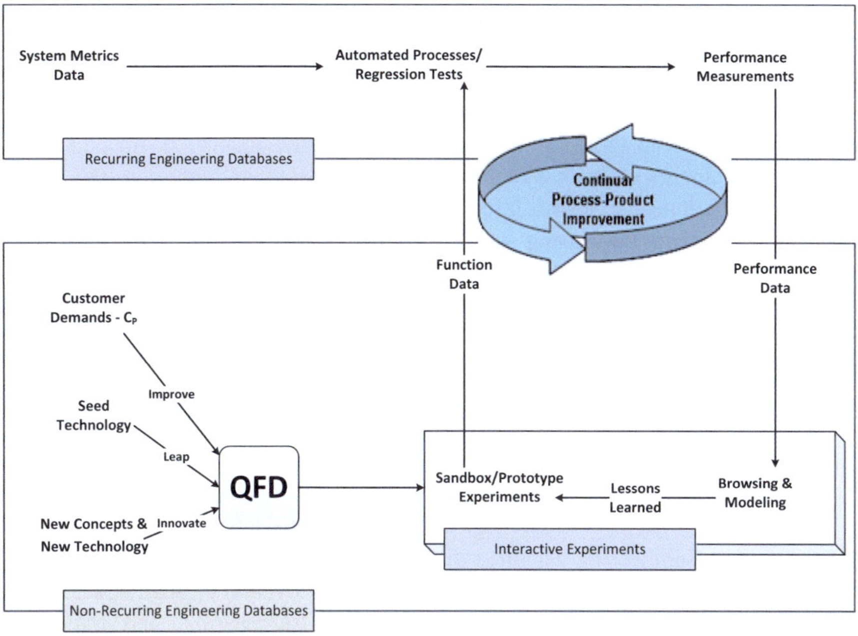

Fig. 11.10 MDSE continual improvement through an integrated approach to engineering

generate the necessary rate of improvement. Treating any one of the domains in isolation will result in ineffectual change. Also, loss of balance between the three will result in loss of competitiveness. Lack of commitment to the three by the organization is a lack of commitment to Quality, People, and Process.

11.5.2.1 Recurring Engineering

Recurring Engineering in enabled through the *Electronic Engineering Notebook* (EEN), and allows software loads to be generated and tested automatically [138]. With all the functions predetermined and all the design tools interfaced through Knowledge Management, the generation of the software loads is simply a matter of CPU time. The process is as follows:

- The analyst completes the data definitions using the data management tools within the EEN.
- Once complete, the engineer can invoke the design macros built into the EEN through the Knowledge Management Process. These macros know how to the read the data and spawn the design processes.
- These processes include the Monte Carlo simulations, and/or regression tests (if required) to measure performance; therefore performance analysis is built into the process that enables requirements verification.

11.5.2.1.1 Interactive Experiments

Interactive Experiments are essential to CPPI as it indicates where the paradigm is weak or broken. Given robust error checking methods and thorough knowledge capture through QFD, there still remains that set of errors that can be traced to inadequate requirements analysis. When addressing this domain, there does not appear to be any substitute for human intuition and insight. At present, the organization depends on peer reviews using viewgraphs and questions. The Expert Designer makes the process-product Functionbases available to anyone over the internal organization web through a variety of collaborative environments, thereby providing familiarity normally reserved for the designer, to the whole community. The least member of the technical community will use the Recurring Engineering functionality and the erudite will use the Interactive Experiments components.

11.5.2.1.2 Non-recurring Engineering

Non-recurring Engineering has three drivers: improvements in *Cp*, leaping to new products (e.g., the Evolving Rapid Prototype), and innovation through new technology and concepts [134]. Making the Recurring Engineering, and Interactive Experiments Functionbases, available to the community will generate an abundance of ideas. The QFD process will manage all these. Each idea to be evaluated against Customer Demands and the impact on Quality Characteristics, then prioritized for new engineering. Discrete engineering reviews will become outmode as the SPC provides progress data real-time. Even the Test Readiness Reviews are overcome by the Software Cleanroom concept for software quality as measurement criteria are established using QFD on day one: if the Recurring Engineering measurements fail minimum requirements, the design in not ready to fly.

11.5.2.1.3 Quality Function Deployment

Quality Function Deployment is employed throughout the life cycle to manage the improvements. Because paradigm paralysis is such a de-motivator to change, an objective basis must be forged to break the deadlock. The QFD process integrates all of the necessary considerations; cost, risk, quality, etc. The list of improvements can be ordered such that the team's resources are always focused on the largest payoff. Diminishing payoff identifies the need for a new paradigm (i.e., new technology and R&D).

11.5.2.1.4 Quality Function Deployment

Continual Process-Product Improvement is a consequence of implementing this paradigm. Quality is not achieved through reacting to failures, it requires both the time and desire to improve the product: automated processes are a prerequisite if time is to be redeemed for new designs; performance measurement is mandatory if quality is to be more than a figment of the managers imagination; browsing and modeling are essential for generation of new ideas and validating existing knowledge; sandbox experiments are the basis for new and novel designs. The whole process is one of "to do, then discover". Improvement and technology development are perpetual.

11.6 Discussion

Change should be an integral component of the MDSE design methodology and affects the whole product and process life-cycle. Because the engineer and his tools are part of this, and because computer technology has had such a widespread impact, a holistic view must be taken. This book is not exhaustive in its coverage of MDSE, but is intended to be an introduction.

The discussion of new organizational structure is a focus on how a change in Systems Engineering to the new MDSE paradigm requires a change in work organization. Engineering organizations were designed when engineering methods were manual and unsophisticated. This new paradigm is possible through the use of modern Information Systems technology and Web Services to manage the mass of information. The real question to be answered is: will the manager cede control to the team? The solution will most likely come through economic necessity, and the sheer inevitability of the Information Age. Whatever happens, a valuable tool in the arsenal is QFD. Engineers can use it to build technology road maps to new products. Managers can use it to build the New Organization to improve Customer Satisfaction.

Increasing Quality should be closely linked to MDSE. Imagine a process with all waste eliminated. Now design it. In contrast, engineering that is driven by conformance-to-spec is suffering paradigm paralysis. To eliminate waste there must be quality in the process; as this will affect quality of the product. For the MDSE design engineer, quality and robustness are often linked, as the measure of performance can be the same for each. Therefore, advanced design algorithms have become even more valued as a means of reducing variance. The toughest challenge for the MDSE may be to make statistics an intuitive skill; solutions to improving quality will then follow more comfortably [135].

Assignments

The philosophy of the assignment development for this textbook was to create a set of though-provoking questions, problems, and design assignments that help drive home the principles, practices, and methodologies of Systems Engineering. We endeavored to provide the opportunity for the student to practice the art and science of Systems Engineering, helping the student to understand that there is rarely one answer to a Systems Engineering problem, and it is only through study and practice that successful Systems Engineers are developed. Systems engineering is a life-long journey, and anyone who tells you they have nothing left to learn doesn't truly understand the discipline.

Overall Design Problem: Edison's Revenge

The purpose of this project is to, over the course of the semester, develop a system specification, derived requirements, and high-level system architecture design for a system that encompasses multiple alternative energy sources and the controlling H/W & S/W to control, store, and distribute DC power to a self-contained 500 home community. Such local power generation systems can be used to power entire communities, while sending the unused energy back to the national power grid.

DC Power Grid Discussion

In the initial days of power generation and distribution, DC power was far less practical than the AC distribution system that soon supplanted it. However, even after Tesla's AC system became the predominant system (and rightly so), Edison's DC power did not become obsolete for some time. Eventually power plant eliminated their delivery of DC power, but the overall concept of a DC power grid is not dead. Even our computer systems (large server systems) have uninterruptible power

J.A. Crowder et al., *Multidisciplinary Systems Engineering*,
DOI 10.1007/978-3-319-22398-8

supplies which contain batteries that are charged off the grid, and used to up-convert to AC to power devices when the AC power grid goes off line. So the notion of a DC power grid is not dead, but generally used as a back-up system. The point is, the notion of a DC power grid has remained and now that renewable energies are viable, we should re-examine their use for widespread, local, power distribution.

One of the advantages of a locally generated DC station is energy savings as well as allowing a facility to run more easily off of various DC sources, such as Solar Arrays. And for power-hungry installations, the notion of generating the power locally is growing in popularity. Part of the reason is that waste heat from the power generation generators (solar, geothermal, etc.) can be used to warm nearby buildings.

By combining the production of heat and power, local power grids can squeeze much more useful energy out of the renewable energy they use, so this approach will certainly become more widespread as time goes on. The purpose of this design problem is to look ahead and provide an architecture and specification for a local DC power generation station that can serve a local community.

Currently many systems exist to control and handle alternative energies. There are systems to control and manage solar power, wind power, geothermal power, and others, but very few are designed to control, manage, store, and distribute energy from a power grid consisting of multiple alternative energy sources.

No one alternative energy source is the answer to the growing energy needs/wants in our society. Each alternative energy source has limitations and issues associated with its use. By harnessing multiple alternative energies, and managing the storage of the energies produces, communities can be worry free from power generation. In order to facilitate utilizing multiple alternative energy sources, a control center must be included in the design that can manage the energy creation, distribution, storage, and return unused energy to the power grid.

Technical Description of the Design Problem

The purpose of the multiple alternative energies DC Power Grid is to:

- Supply DC power to a customer community of 500 homes and a biosphere for growing food.
- Provide a storage device to store energy in case of a system failure.
- Determine the amount of surplus energy that is above and beyond what is required to power the grid and charge the backup system, and then deliver that surplus energy to the national power grid systems.

Throughout the semester you will be asked to analyze, design, and architect different components of the overall DC power grid system. This will include research into alternative designs, like Lithium Nanowire Storage systems for backup power and/or Super Capacitors (also called Ultracapacitors). Figure A.1 illustrates these technologies connected to the DC Power Grid Regulator Control Center.

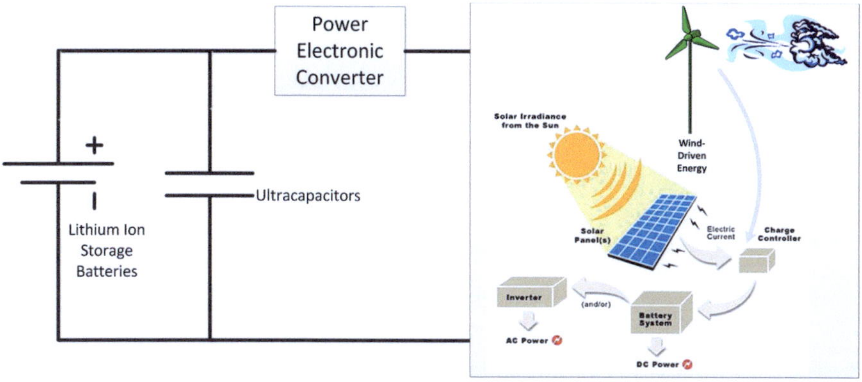

Fig. A.1 Alternative power management and control center with power storage devices

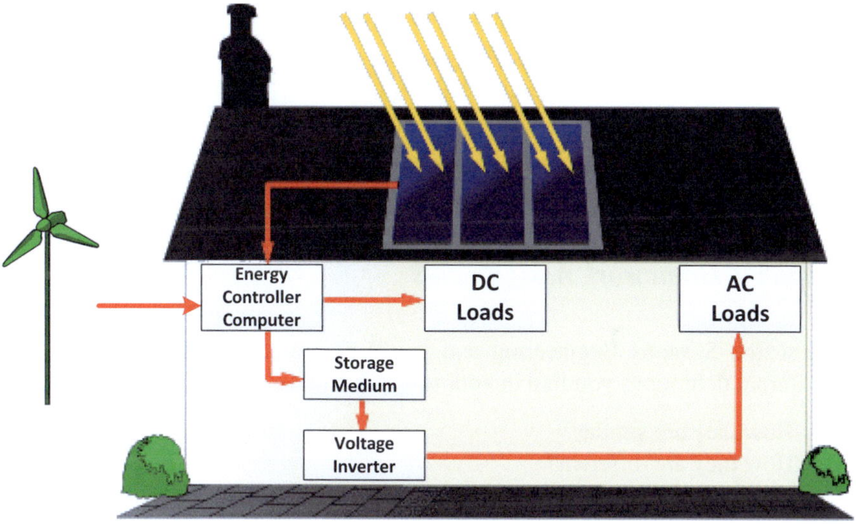

Fig. A.2 Whole house power grid

Technical Requirements for the Design Problem

- Design shall utilize at least two alternative energies, with a maximum of four.
- Total Power Generation shall be 3.0 MW minimum
- Each of the alternative energy sources shall be capable of generating at least 60 % of minimum wattage.
- DC Power Grid shall generate DC power, with Grid voltage being 48, 24, or 12 V
- DC Power Grid shall be capable of servicing at least 500 single family homes and one 4-acre biosphere with hydroponics capable of growing vegetables year-round. (See Fig. A.2 for a block diagram of a DC-powered home)

- All lights, appliances, and facilities in each home and the biosphere shall be configured for DC operations.
- Only allowed exceptions are the HVAC system and the Washer/Dryer, although DC is preferred.
- There shall be energy storage capability capable of operating the grid of 500 single family dwellings for 7 days if all alternative forms of energy are not producing.
  - Compliance Criteria: Energy Storage Devices shall be capable of producing 750,000 W for the entire 7 day period. Nominal Energy Storage should be Batteries or Supercapacitors.
- The DC Power Grid System shall include controlling h/w and s/w to manage the multi-alternative energy strategies and production.
- All energy produced above the required home and energy storage usage shall be up-converted to 220vac for release to the national power grid.
- The DC Power Grid shall be capable of operating in a temperature range of −40 to 125 °F without degradation of power generation.
- The DC Power Grid must meet OSHA regulations

Chapter 1 Assignments

Chapter 1 Homework Assignments

- Research Systems Engineering and you'll find many definitions. Pick three different definitions you find in your research and describe:
  - How they are similar?
  - How they are different?
  - How might the systems designs be different if each of the definitions were utilized?
- What role does Systems Engineering play in the overall success of modern engineering systems?
- How do you believe Systems Engineering enhances the development of advanced Communications and Information Systems?
- How does a company or organization's overall business/mission strategy affect its Systems Engineering organization?
- What is an Enterprise Architecture and why is it important?
- Do you believe Cognitive Sciences is required for MDSE? Why or Why not?
- How important do you feel Human-Systems Engineering is to the overall success of a system design?

Chapter 1 Design Assignments

Spend time researching alternative, renewable energy and energy storage technologies, based on the description in Sect. 1.7. Write a paper providing your decisions about the following:

1. Which alternative energy and energy storage technologies are you going to utilize in your DC Grid design project?
   a. Remember you must choose at least two and no more than four.
2. What features drew you to the alternative energies you chose?
3. Why did you choose the energy storage technologies you are going to use?
4. Show the calculations on how you are going to generate the required power from your DC grid, based on the requirements given in Sect. 1.7.2.
5. Which engineering and other disciplines must you embrace in order to design the DC Grid, as well as the Biosphere?

Chapter 2 Assignments

Chapter 2 Homework Assignments

- In order to develop keen insight into the First Principles, pick an MDSE discipline from Fig. 1.5. Describe in detail and defend at least three core First Principles of the chosen domain.
- Refer to Fig. 2.1. Do you feel the current Science, Technology, Engineering, and Math curriculums in Middle and High Schools adequately prepare students for a curriculum in Multidisciplinary Systems Engineering? Defend your position.
- Referring to Sect. 2.5. Why do you think engineers, who are supposed to live in an environment of change, are, themselves, so reluctant to change?
- Cyber Security has become a major curriculum and discipline of its own. How important is it for the Multidisciplinary Systems Engineer to understand and embrace Information Assurance, particularly when it applies to Data/Information Security.
- Why do you think a common lexicon is so important within the overall System of Systems design?

Chapter 2 Design Assignments

1 Looking at the high-level systems requirements for the Design Problem, identify three major risks you can envision in the development of the DC Power Grid.

2 Do you feel the DC Power Grid design problem qualifies as a System of Systems? Justify your answer, as this will drive your overall design architecture.

3 Using the high-level systems requirements for the Design Problem, identify the major system functions for the DC Power Grid.

4 Define the major system functions for the Biosphere element of the overall System of Systems DC Power Grid project.

Chapter 3 Assignments

Chapter 3 Homework Assignments

- Review the description of the Systems Designer and answer the following questions:
  - Do you believe the Systems Architecture/Analyst steps can be eliminated and go right to the Systems Designer work? Why or Why Not?
  - Research bottoms-up design. When would a bottoms-up approach to design be desirable over a top-down approach?
- Review the description of the ATAM methodology and answer the following questions:
  - Why might it not be possible to meet all of the quality attributes completely?
  - Explain the need for the ATAM Sensitivity Analysis
- Research the IEEE 29148-2011 Standard and the IEEE 12207 Standards and answer the following questions:
  - Which of the architectural frameworks discussed in Chap. 4 do you feel comply best with these standards documents? Please justify your answer.
  - What are the differences between the EIA-632 and the IEEE 29148-2011 standards?
  - What is the overall mission of INCOSE? Do the INCOSE standards of complex system conceptualization provide compare to DoDAF 2.2 in terms of Data-Centric System of Systems Architectural design?
- Review the Case Study on Ontologies and answer the following questions:
  - Do you believe ontology development is useful for Knowledge Management?
  - What affect would there be on the SoS development if Knowledge Management is not part of the overall Systems Architecture?
  - Why do you think a Fault Ontology is required for effective system design and operations?

Chapter 3 Design Assignments

1. Using the major system functions you created in the Chap. 2 design assignment, decide how to allocate the high-level system requirements into these system functions. If a given requirement does not completely map into one system function, decompose it by deriving two or more requirements (depending on how many functions it maps into), based on the system-level requirement, which can be completely allocated into each system function.

2. Using the description of the DC Grid and the high-level system requirements write two usage scenarios based on utilizing the DC grid to power a 500 home community.
3. Write a three page paper on how data will be managed in your DC Power Grid system.

Chapter 4 Assignments

Chapter 4 Homework Assignments

- Research the DoDAF, TOGAF, and MODAF architecture frameworks discussed above and answer the following questions:
  - Which framework provides the best overall architectural framework with which to create a System of Systems Architecture?
- Review the Zachman Framework and answer the following questions:
  - Why is the Contextual Model row important to the overall system architecture design?
  - What do you think would happen to the overall success of the program if the Conceptual Model row was disregarded during the systems architecture development?
  - Can any of the architectural frameworks methodologies discussed in Chap. 4 fit into the Zachman Framework Matrix?
- Review the Case Study and answer the following questions:
  - What effect did skipping the interface test have on the success of the satellite launch?
  - Could this incident have been prevented a different way other than through an efficient Configuration Management Process?

Chapter 4 Design Assignments

1. Research the architecture frameworks discussed in Chap. 3. Which architecture framework will you used for the DC Power Grid design project and why?
2. For each of your element-level functional blocks created in Chap. 2 and the decomposed/derived/allocated requirements from Chap. 3, decompose each functional block one level down, creating sub-functional blocks for each major system function and decompose/derive/allocate the requirement allocated to each element level. Explain why you chose to decompose each element-level functional block the way you have shown it.
3. Write two usage scenarios (Use Cases) on how you envision the Biosphere to be utilized within the community powered by your DC Grid.

Chapter 5 Assignments

Chapter 5 Homework Assignments

- Describe the difference between functional, non-functional, and performance requirements.
- How do Quality Attributes affect the overall system design? Give at least three examples and explain.
- Of the requirements decomposition/derivation guidelines describe in Sect. 5.1.1, which two do you feel are the most difficult to comply with and why?
- Why is Designing for Maintenance so important?
- Review the Integration Requirements in Sect. 5.4. What problems can you envision for overall system operations if the system is not designed for System of Systems element integration?
- Describe, in your own words, the concept of operational suitability. Why do you feel this is important for the up-front architectural design?
- Review the Case Study for Chap. 5. What do you think the effect will be on companies that were relying on IBM Blade Servers for their IT infrastructure?

Chapter 5 Design Assignments

1. Derive the technical requirements, based on the high-level requirements given, for the alternative energies you have chosen for your design project.
2. Derive the functional requirements based on the high-level requirements given, for the alternative energies you have chosen for your design project.
3. Define what you believe are applicable Reliability, Availability, and Maintainability requirements for your DC Power Grid design.

Chapter 6 Assignments

Chapter 6 Homework Assignments

- Describe the difference between formal and informal requirements.
- Describe problems that informal requirements may pose to the overall system design.
- Research the concept of Operations Logging. Why is this essential to overall system maintainability?
- What are the differences between Use Cases, Sequence Diagrams, and Activity Diagrams? What information does each provide that is not provided in the other two?

- Review the Case Study for Chap. 6. How do you believe the program organizational structure can help/hurt an overall program success/failure?

Chapter 6 Design Assignments

1. Consider the possible failure modes of your DC Power Grid System. Describe two possible/probable failure modes and how you will mitigate these within your system.
2. Define two Use Scenarios for you DC Power Grid System. Use these scenarios to create Use Cases and Activity Diagrams (for this exercise, PowerPoint is ok, unless you are directed otherwise).
3. Define three major system tests that will be required for your DC Power Grid.

Chapter 7 Assignments

Chapter 7 Homework Assignments

- Why are formal reviews essential to the success of a program, particularly large programs?
- How can the mapping of Parent/Child requirements either enhance system development if done correctly, or hamper system development if done incorrectly?
- Describe the general rules for requirements attributes.
- Research Tiered architectures. Describe a Presentation Tier and why it is an essential element of any system that includes human operators.
- Describe a Trade Study and explain its use in the overall architecture design process.

Chapter 7 Design Assignments

1. Describe two Trade Studies that should be done for your DC Grid design project.
2. Describe one internal and one external interface that will be required for your DC Grid design.
3. Decompose your system to the next level (Subsystem Level) and decompose/derive the requirements needed to describe your Subsystem Level.
4. Describe three factors that will affect the overall Lifecycle Costs for your DC Power Grid System.
5. Describe three major risks associated with your DC Power Grid System. How would you provide mitigation for these risks?

Chapter 8 Assignments

Chapter 8 Homework Assignments

- Why do you think creation of a SEMP is important?
- What are the potential problem associated with not having an Integrated Verification and Validation Plan?
- Describe the difference between verification and validation. Is it reasonable to do one without the other? Explain your answer.
- Why is Configuration Management so important throughout the system development life cycle
- Research the Agile Software Development methodology. Does agile software development still require a software development plan? Why or why not?
- What is an Integrated Test Plan? What are the ramifications of not adequately testing the overall integrated system?
- What is Information Assurance? Why is it important for Information Security to be a part of the overall system architecture?
- What could happen if a Facilities Plan is not developed before the system is deployed?

Chapter 8 Design Assignments

1. Create a System Breakdown Structure for you DC Power Grid Design.
2. Describe four aspect of system safety that would be important in your DC Power Grid system.
3. Describe the ramifications of hackers gaining control of your DC Power Grid system.
4. What type of training manuals do you think you would need to provide to the maintainers of your DC Power Grid System?
5. Create a Work Breakdown Structure for your DC Power Grid System
6. Describe four main system functions that would need to be Validated and Verified for you DC Power Grid system.
7. Research Reliability and Availability Quality attributes. Define the reliability and maintainability you believe is required for your DC Power Grid system.

Chapter 9 Assignments

Chapter 9 Homework Assignments

- Describe the need for a formal "Authorization to Proceed" milestone, even if the program is internal to a company.

- Define top-down vs. bottom-up integration. What are the advantages and disadvantages of each?
- Why are Interface Control Documents (ICDs) necessary for external interfaces used of the SoS and what role do they play?
- What problems can you envision if the Final Design Review left out of the overall MDSE process?
- Do you feel it's possible for the systems and software architectures to both meet their requirements, but not be in sync with each other? Describe the problems with them not being in sync.
- Describe Data management and its role within the SoS operations and management.
- Describe the potential issues that may ensue if more than one subsystem utilizes the same Configuration Item.
- Why is data integrity important?
- Why is it important to provide Data Governance within the SoS architecture?

Chapter 9 Design Assignments

1. Define four software services that would be required in the development of your DC Grid system.
2. Define and describe four Configuration Items that will be required for you DC Grid system.
3. Which Integration and Test methodology (top-down, bottom-up, or hybrid) would you envision is best for your DC Grid system?
4. Define four Data Items (DIs) that will be required for your DC Power Grid system.

Chapter 10 Assignments

Chapter 10 Homework Assignments

Please review the Chap. 10 Case Study and answer the following questions.

System Acquisition Assessment

- What customer actions introduced failures into their acquisition,
    - For the CDP and CPP phases?
    - For the Program startup phase?
    - For the period of performance to SRR?
    - For the period of performance to PDR1?

  - For the period during the first Stop Work Period?
  - For the period of performance to PDR2?
  - For the second restart period?

- When should the first stop work order have been issues?
- When should the second stop work order have been issues? (hint: its exposed in the second PDR period)
- Should the second restart been undertaken, and what actions should have been taken by the customer?
- Since the customer did initiate the second restart, what were the factors that lead to the death knell for program?

Vendor Development Assessment

- What vendor actions introduced failures in the development,
  - For the CDP and CPP phases?
  - For the Program startup phase
  - For the period of performance to SRR
  - For the period of performance to PDR1
  - For the period during the first Stop Work Period
  - For the period of performance to PDR2
  - For the second restart period
- From an MDSE point of view,
  - What actions should the Prime have taken right after contract startup?
  - What actions should the Prime have taken leading into SRR?
  - What actions should have been taken leading to PDR #2?
- Since the CONOPS, Requirements, Architecture and Transition were acceptable at re-start,
  - What action should the Prime have taken when new requirements were levied on them?
- Since the program went to stop work for a second time,
  - What action should the Prime have taken when a fourth vendor was added to the mix?, and why?

Chapter 10 Design Assignments

1. Define the elements of your DC Power Grid system.
2. Create a Concept of Operations (CONOPS) for your DC Power Grid System (5 pages).

3. Decompose/Derive the next level of requirements for your DC Power Grid System.
4. Which architecture views do you feel are required for your DC Power Grid System?
5. Write five Test Cases for your DC Power Grid System.
6. Describe the User Interfaces you envision for your DC Power Grid system.

Chapter 11 Assignments

Chapter 11 Homework Assignments

- Do you believe reuse is a realizable concept in large System of Systems? Justify your answer.
- Define "Reactionary Engineering." When is this appropriate?
- Define Agile Systems Engineering. Why is it necessary for complex System of Systems developments?
- Describe the difference between data, information, and knowledge.
- What are the pitfalls of using information from the Web?
- How would automation help/hurt System of Systems Integration and Test?
- How can Functionbases be used to capture the context of a system design?

Chapter 11 Design Assignments

1. Create a 25 slide presentation of your DC Grid project design.
2. What MDSE processes have you used?
3. What MDSE processes did you feel were not necessary for your DC Grid system?
4. Do you believe your DC Grid design is viable? Justify your answer.

Acronyms

ACT-R	Adaptive Control of Thought-Rational
ADM	Architecture Development Method
AI	Artificial Intelligence
API	Application Programming Interface
ASCII	American Standard Code for Information Interchange
ATAM	Architecture Tradeoff Analysis Method
ATP	Authorization to Proceed
CCD	Charge Coupled Device
CDP	Concept Definition Phase
CERN	Council Europeen pour la Recherche Nucleaire
CI	Configuration Item
CM	Configuration Management
CMMI	Capability Maturity Model Integration
CMP	Configuration Management Plan
COGNET	Cognitive Network
CONOPS	Concept of Operations
COTS	Commercial off the Shelf
CPP	Concept Prototype Phase
CPPI	Continuous Product/Process Improvement
CPU	Central Processing Unit
CSCI	Computer Software Configuration Item
CT	Computational Thinking
CUT	Code and Unit Test
DAO	Data Access Objects
DBM	Database Management
DCID	Director of Central Intelligence Directive
DFM	Designing for Maintenance
DI	Data Item
DIA	Denver International Airport

J.A. Crowder et al., *Multidisciplinary Systems Engineering*,
DOI 10.1007/978-3-319-22398-8

DID	Date Item Description
DMZ	Demilitarized Zone
DNA	Deoxyribonucleic Acid
DoD	Department of Defense
DoDAF	Department of Defense Architecture Framework
DPI	Dots Per Inch
DR	Deficiency Report
DTED	Digital Terrain Elevation Data
EDM	Enterprise Data Management
EEN	Electronic Engineering Notebook
EEOC	Equal Employment Opportunity Commission
EIA	Engineering Industries Alliance
EPIC	Executive Process—Interactive Control
ERM	Enterprise Risk Management
FAA	Federal Aviation Administration
FCC	Federal Communications Commission
FDR	Final Design Review
FMEA	Failure Modes and Effects Analysis
FOSS	Free and Open Source Software
GPS	Global Positioning System
GUI	Graphical User Interface
HCI	Human-Computer Interface
HDP	Hardware Development Plan
HVAC	Heating, Ventilation, and Air Conditioning
HW	Hardware
I/O	Input/Output
IA	Information Assurance
IBM	International Business Machine
ICD	Interface Control Document
ID	Intelligent Design
IDEAS	International Defense Architecture Specification
IDLS	Ideally Diffuse Light Source
IE	Industrial Engineers
IEEE	Institute of Electrical and Electronics Engineers
IMP	Integrated Master Plan
IMS	Integrated Master Schedule
INCOSE	International Counsel of Systems Engineering
IoT	Internet of Things
IP	Internet Protocol
IRS	Interface Requirement Specification
ITP	Integration and Test Plan
LCC	Life Cycle Costs
LHC	Large Hadron Collider
LoO	Likelihood of Occurrence
LRU	Line Replaceable Unit

M&S	Modeling and Simulation
MDSE	Multidisciplinary Systems Engineering
MODAF	Ministry of Defense Architecture Framework
MODEM	MODAF Ontological Data Exchange Mechanism
MOE	Measures of Effectiveness
MTTR	Mean Time to Repair
NASA	National Aeronautics and Space Administration
NATO	North Atlantic Treaty Organization
O&S	Operations and Sustainment
OBS	Organizational Breakdown Structure
OCE	Office of Chief Engineer
OOAD	Object Oriented Analysis and Design
OOD	Object Oriented Design
OSHA	Occupational Safety and Health Administration
OV	Operational View
PC	Personal Computer
PDR	Preliminary Design Review
PMP	Program Management Plan
PPS&C	Programming Practices, Standards, and Conventions
QA	Quality Assurance
QFD	Quality Functional Deployment
RDBMS	Relational Database Management System
RF	Radio Frequency
RMA	Reliability, Maintainability, Availability
RNA	Recombinant kNowledge Assimilation
SAP	Site Activation Plan
SBS	System Breakdown Structure
SDP	Software Development Plan
SE	Systems Engineering
SELF	Self-Evolving Life Form
SEMP	Systems Engineering Management Plan
SEP	Systems Engineering Plan
SI	Systems Integration
SLA	Service Level Agreement
SLOC	Software Lines of Code
SOA	Service Oriented Architecture
SOAR	State, Operator, and Result
SoO	Statement of Objectives
SoS	System of Systems
SOW	Statement of Work
SQA	Software Quality Assurance
SRR	System Requirements Review
STEM	Science Tech
SV	System View
SW	Software Quality Assurance

TBD	To Be Determined
TCO	Total Cost of Ownership
TCP/IP	Transmission Control Protocol/Internet Protocol
TDD	Test-Driven Design
TEMP	Test Engineering Management Plan
TOGAF	The Open Group Architecture Framework
TPM	Technical Performance Measure
UAV	Unmanned Air Vehicle
UDP	User Datagram Protocol
UL	United Laboratories
UML	Unified Modeling Language
USAF	United States Air Force
USB	Universal Serial Bus
USG	United States Government
USN	United States Navy
V&V	Verification and Validation
WBS	Work Breakdown Structure
XML	Extensible Markup Language

References

1. Creswell, J. 2003. Research Design: Qualitative, Quantitative and Mixed Approached. Sage.
2. Camarinha-Matos, L. and Afsarmanesh, H. 2008. Collaborative Networks: Reference Modeling. Springer Publishing, New York NY.
3. Clements, P. 1996. A survey of architecture description languages. Proceedings of the 8th international workshop on software specification and design. IEEE Computer Society.
4. Booch, G., Rumbaugh, J., and Jacobson, I. 1998. The Unified Modeling Language Users Guide. Addison Wesley Publishing, Boson, MA, ISBN: 0-201-57168-4.
5. Kruchten, P. 2000. The rational unified process: an introduction. Addison-Wesley Longman Publishing Co., Inc. Boston, MA.
6. Kossiakoff, A. and Sweet, W. 2003. Systems Engineering: Principles and Practice. John Wiley & Sons, Hoboken, NJ.
7. Luzeaux, D., and Rualt, J. 2010. Systems of Systems. ISTE Ltd and John Wiley & Sons Inc., New York, NY.
8. Ross, J., Weill, P., and Robertson, D. 2006. Enterprise Architecture as Strategy: Creating a Foundation for Business Execution. Harvard Business Review Press, Harvard University, Cambridge, MA.
9. Bass, L., Clements, P., and Kazman, R. 2003. Software Architecture in Practice, Second Edition. Addison Wesley Professional, Boston, MA.
10. Doerr, J., Kerlow, D., Koenig, T., Olson, T., and Suzuki, T. 2005. Non-Functional Requirements in Industry – Three Case Studies Adopting the ASPIRE NFR Method Technical Report 025.05/E, Fraunhofer IESE.
11. Parnas, D. 1979. Designing Software for ease of Extension and Contraction. IEEE Transactions of Software Engineering, 5(2):128–138.
12. Parker, J. 2010. Applying a System of Systems Approach for improved transportation. S.A.P.I.EN.S. 3 (2).
13. Simon, D. 2000. Design and Rule Base Reduction of a Fuzzy Filter for the Estimation of Motor Currents. International Journal of Approximate Reasoning, 25(2).
14. Maier, M. 1998. Architecting Principles for System of Systems. Systems Engineering 1 (4): 267–284.
15. Egyed, A., Grunbacher, P., and Medvidovic, N. 2001. Refinement and Evolution Issues in Bridging Requirements and Architecture-the CBSP Approach. In International Software Requirements to Architecture Workshop, Toronto, CAN.
16. Bahrami, A. 1999. Object-Oriented Systems Development: Using the Unified Modeling Language. McGraw Hill, New York, NY, ISBN 0-07-234966-2.

J.A. Crowder et al., *Multidisciplinary Systems Engineering*,
DOI 10.1007/978-3-319-22398-8

17. Crowder, J. and Carbone, J. 2012. Reasoning Frameworks for Autonomous Systems. Proceedings of the AIAA Infotech@Aerospace 2012 Conference, Garden Grove, CA.
18. Jamshidi, M. 2005. System-of-Systems Engineering - A Definition. IEEE SMC 2005, 10–12.
19. Conway, W. 1995. The Quality Secret: The Right Way to Manage. Conway Quality, Nashua, NH.
20. Books, F. 1975. The Mythical Man Month. Addison Wesley, Boston, MA.
21. Beck, K; et al. 2001. Manifesto for Agile Software Development. Agile Alliance.
22. Alpern, B. and Schneider, F. 1987. Recognizing Safety and Liveness. Distributed computing, 2(3):117–126.
23. Abadi, M. and Lamport, L. 1991. The Existence of Refinement Mappings. Theoretical Computer Science, 82(2):253 – 284.
24. Cheng, B. and Atlee, J. 2009. Current and Future Research Directions in Requirements Engineering. In Design Requirements Engineering: A Ten-Year Perspective Workshop, Cleveland, OH.
25. Tolk, A. and Lakhmi, J. (Eds.). 2011. Intelligence-Based Systems Engineering. Springer Publishing, New York, NY. ISBN 978-3-642-17930-3.
26. Manthorpe, W. 1996. The Emerging Joint System-of-Systems: A Systems Engineering Challenge and Opportunity for APL. Johns Hopkins APL Technical Digest, Vol. 17, No. 3, pp. 305–310.
27. Ertas, A. and Tanik, M. 2000. Transdisciplinary Engineering Education and Research Model. Transactions of the SDPS, Vol. 4, No. 4, pp. 1-11.
28. Morris, P., Hough, G., and Morris, W. 1987. The Anatomy of Major Projects: A Study of the Reality of Project Management. Wiley & Sons, Chichester, UK.
29. Feathers, M. 2004. Working Effectively with Legacy Code, Prentice Hall, New York, NY.
30. Yaneer, B. 2004. The Characteristics and Emerging Behaviors of System-of-Systems. In: NECSI: Complex Physical, Biological and Social Systems Project.
31. Medvidovic, N., Grunbacher, P., Egyed, A., and Boehm, B. 2003. Bridging Models across the Software Lifecycle. Journal on Systems and Software, 68:3.
32. Mittal, S. and Martin, J. 2013. Netcentric System of Systems Engineering with DEVS Unified Process. CRC Press, Boca Raton, FL.
33. Curry, E. 2012. System of Systems Information Interoperability Using a Linked Dataspace. In IEEE 7th International Conference on System of Systems Engineering (SOSE 2012), 101–106.
34. Lucena, J. 2013. Engineering Education for Social Justice. Springer Publishing, New York, NY, ISBN 978-94-007-6350-0.
35. McHugh, O. Conboy, K., and Lang, M. 2012. Agile Practices: The Impact on Trust in Software Project Teams. IEEE Software May/June 2012.
36. Crawford, B., Leon de la Barra, C., Soto, R., and Monfroy, E. 2012. Agile Software Engineering as Creative Work. Proceedings of the 5th International Workshop on Co-operative and Human Aspects of Software Engineering, ICSE, Zurich, Switzerland.
37. Azim, S., Gale, A., Lawlor-Wright, T., Kirkham, R., Khan, A., and Mehmood, A. 2010. The Importance of Soft Skills in Complex Projects. International Journal of Managing Projects in Business, 3, 387–401.
38. Meier, B., Wilkowski, B., and Robinson, M. 2007. Bringing out the Agreeableness in Everyone: Using a Cognitive Self-Regulation Model to Reduce Aggression. Journal of Experimental Social Psychology, 44(2008), 1383–1387.
39. Wingreen, N. and Botstein, D. 2006. Back to the Future: Education for Systems-Level Biologists. Nature Reviews Molecular Biology, 7, 829–832.
40. Antonsson, E. 1987. Development and Testing of Hypotheses in Engineering Design Research. ASME Journal of Mechanisms, Transmissions, and Automation in Design. Vol. 109(2), 153–154.
41. Medvidovic, N. and Taylor, R. 2000. A classification and comparison framework for software architecture description languages. IEEE Transactions on Software Engineering 26.1: 70–93.

42. Schenk, T. 2005. Introduction to Photogrammetry. Department of Civil and Environmental Engineering and Geodetic Science, Ohio State University, Columbus, OH.
43. Hartman, J. 2006. A Systems Approach to Understanding Gene Interaction Networks using the Yeast Model System. SDPS Systems Biology Workshop, San Diego, CA.
44. Landis, R., Carbone, J, and Tosetta, C. 2007. Transdisciplinary Approaches to Systems Biology. Curriculum for Texas Tech ME6331, Systems Biology for Engineers.
45. Willson, R. 1994. Modeling and Calibration of Automated Zoom Lenses. In SPIE 2350:Videometrics III, Boston, MA.
46. Shan, T. and Hua, W. 2006. Solution Architecting Mechanism. Proceedings of the 10th IEEE International EDOC Enterprise Computing Conference (EDOC 200), pp. 23–32.
47. Wang, Ostermann, and Zhang. 2001. Video Processing and Communications. Prentice Hall Publishing, Upper Saddle River, NJ.
48. Willson, R. and Schafer, S. 1993. A Perspective Projection Camera Model for Zoom Lenses. In Proceedings of the Second Conference on Optical 3-D Measurement Techniques, Zurich, Switzerland.
49. Kjokic, S. 2003. The Moving Optical Center in Lighting Calculations. Journal of the Illuminating Engineering Society, Vol. 32.
50. Weghorst, H., Hooper, G., and Greenberg, D. 1984. Improved Computational Methods for Ray Tracing. ACM Transaction of Graphics, Vol. 3.
51. Curcin, V., Ghanem, M., Guo, Y., and Rowe, A. 2004. IT Service Infrastructure for Integrative Systems Biology. In IEEE International Conference on Services Computing.
52. Fiadeiro, J., Lopes, A., and Bocchi, L. 2006. A Formal Approach to Service Component Architecture. Web Services and Formal Methods, 4148:193–213.
53. Cheng, J., Scharenbroich, L., Baldi, P., and Mjolsness, E. 2005. Sigmoid: A Software Infrastructure for Pathway Bioinformatics and Systems Biology. IEEE Intelligent Systems.
54. Shah, N., Laws, J., Wardman, B., Zhao, P., and Hartman, L. 2007. Accurate, Precise Modeling of Cell Proliferation Kinetics from Time-Lapse Imaging and Automated Image Analysis of Agar Yeast Culture Arrays. BMC Systems Biology, Vol. 1.
55. Carnegie Mellon. 2013. Architecture Tradeoff Analysis Method. Carnegie Mellon Software Engineering Institute, Pittsburg, PA.
56. Crowder, J. and Friess, S. 2013. Systems Engineering Agile Design Methodologies. Springer Publishing, New York, NY. ISBN-10: 1461466628.
57. Egyed, A., Grunbacher, P., Heindl, M., and Biffl, S. 2009. Value-Based Requirements Traceability: Lessons Learned. Design Requirements Engineering: A Ten-Year Perspective. Springer-Verlag, Berlin Heidelberg, GER, ISBN 978-3-540-92966-6.
58. Safonov, M. and Laub, A. 2001. The Role and Use of the QFD Matrix. IEEE Transactions in Automatic Control.
59. Burgin, M. 1982. Generalized Kolmogorov complexity and duality in theory of computations. Notices of the Russian Academy of Sciences, v.25, No. 3, pp. 19–23.
60. Pavlich-Mariscal, J. 2005. A Framework for Composable Security Definition, Assurance, and Enforcement. MODELS Conference, Ottawa, CAN.
61. Bailey, I. and Partridge, C. 2009. Working with Extensional Ontology for Defense Applications. Ontology in Intelligence Conference, GMU, Fairfax, VA.
62. Crowder, J. 2003b. Ontology-Based Knowledge Management. NSA Technical Paper – CON-SP-0014-2003-05.
63. Hershey, P. and Blanchard, K. 1989. Management of Organizational Behavior: Utilizing Human Resources. Prentice-Hall, Saddle River, NJ.
64. Offutt, J. and J. Hayes. 1996. A Semantic Model of Program Faults. International Symposium on Software Testing and Analysis. CiteSeerX: 10.1.1.134.9338.
65. Meyers, R. 2009. Encyclopedia of Complexity and Systems Science. Springer Publishing, New York, NY, ISBN 978-0-387-75888-6
66. Kotonya, G., and Sommerville, I. 1998. Requirements Engineering: Processes and Techniques. John Wiley & Sons Ltd, Hoboken, NJ, ISBN: 978-0-471-97208-2.

67. Zachman, J. 2003. The Zachman Framework For Enterprise Architecture: Primer for Enterprise Engineering and Manufacturing. Zachman International.
68. Zachman, J. 1987. A Framework for Information Systems Architecture. IBM Systems Journal, Volume 26, Number 3.
69. Reynard, W., Billings, C., Cheaney, E., and Hardy, R. 1986. The Development of the NASA Aviation Safety Reporting System. NASA Reference Publication 1114.
70. DeLaurentis, D. 2007. System of Systems Definition and Vocabulary. School of Aeronautics and Astronautics, Purdue University, West Lafayette, IN.
71. Helmreich, R., and Wilhelm, J. 1991. Outcomes of Crew Resource Management training. International Journal of Aviation Psychology, 1(4), 287–300.
72. Chonoles, M. and Schardt, J. 2003. UML for Dummies. Wiley Publishing, Hoboken, NJ, ISBN 0-7645-2614-6.
73. Pender, T. 2003. UML Bible. Google Books, ISBN 0-7645-2604-9.
74. Penzenstadler, B. and Koss, D. 2008. High Confidence Subsystem Modeling for Reuse. International Conference on Software Reuse, Beijing, China.
75. Clausing, D. 2001. Concurrent Engineering. Design and Productivity International Conference.
76. Heinz, J. 2000. What Went Wrong. Aerospace & Defense Science.
77. Heidrick, J., Munch, J., Riddle, W., and Rombach, D. 2006. New Trends in Software Process Modelling. World Scientific, Volume 18, ISBN: 978-981-256-619-5.
78. Pohl, K. and Sikora, E. 2007. Structuring the Co-Design of Requirements and Architecture. International Conference on Requirements Engineering: Foundations for Software Quality. Springer Publishing, New York, NY.
79. Chung, L., Nixon, B., Yu, E., and Mylopoulos, J. 2000. Non-Functional Requirements in Software Engineering. Kluwer Academic Publishers.
80. Paech, B., et al. 2003. An Experience-Based Approach for Integrating Architecture and Requirements Engineering. International Software Requirements to Architecture Workshop, STRAW 2003, Portland Oregon.
81. Goteland, O. and Finkelstein, C. 1994. An Analysis of the Requirements Traceability Problem. International Conference on Requirements Engineering, Colorado Springs, CO.
82. Korel, G. 1990. Automated Software Test Data Generation. IEEE Transactions on Software Engineering, CiteSeerX: 10.1.1.121.8803.
83. Copeland, L. 2001. Extreme Programming. Computerworld.
84. Koskela, L. 2007. Test Driven: TDD and Acceptance TDD for Java Developers. Manning Publications, Shelter Island, NY.
85. Beck, K. 2003. Test-Driven Development by Example. Addison Wesley – Vaseem, Boston, MA.
86. Beck, K. 2002. Test Driven Development. Addison-Wesley Professional, Boston, MA.
87. Boehm, B. 1984. Verifying and Validating Software Requirements and Design Specifications. IEEE Software, 1(1):75–88.
88. Perks, C., and Beveridge, T. 2003. Guide to Enterprise IT Architecture. Springer Publishing, New York, NY, ISBN: 0-387-95132-6
89. Manfred, B. 2007. Model-driven architecture-centric engineering of (embedded) software intensive systems: modeling theories and architectural milestones. Innovations in Systems and Software Engineering, 3(1):75–102.
90. Bendat, J. 2000. Non-Linear System Analysis and Identification. Wiley Interscience, New York, NY.
91. Beck, K. 1999. XP Explained, 1st Edition. Addison-Wesley Professional, Boston, MA ISBN 0201616416.
92. Herrmann, A., Paech, B., and Plaza, D. 2006. An Automated Process for the Solution of Requirements Conflicts and Architecture Design. International Journal and Knowledge Engineering, 16:1–34.
93. Petroski, H. 1992. To Engineer is Human: The Role of Failure in Successful Design. Vintage Books, New York, NY, ISBN-10: 0679734163.

94. Russell, S. and Zilverstein, S. 1991. Composing Real-Time Systems. In Proceedings of the Twelfth International Joint Conference on Artificial Intelligence, Sydney, Australia.
95. Dwyer, M., Avrunin, G., and Corbett, J. 1999. Patterns in Property Specifications for Finite-State Verification. In Proceedings of the International Conference on Software Engineering, Los Angeles, CA.
96. Ramesh, B. and Jarke, M. 2001. Toward Reference Models for Requirements Traceability. IEEE Transactions on Software Engineering 27(1):58–93.
97. Robertson, J. and Robertson, S. 2007. Mastering the Requirements Process. Addison-Wesley, Boston, MA, ISBN-13: 978-0321815743.
98. Tyree, J. and Akerman, A. 2005. Architecture Decisions: Demystifying Architecture. IEEE Software, 22(2):19–27.
99. Lamsweerde, A. 2001. Goal-Oriented Requirements Engineering. In Proceedings of the Fifth IEEE International Symposium on Requirements Engineering, Toronto, CAN.
100. Lamsweerde, A. 2009. Requirements Engineering: From System Goals to UML Models to Software Specifications. Wiley & Sons, Hoboken, NJ.
101. Broy, M. and Stølen, K. 2001. Specification and Development of Interactive Systems: Focus on Streams, Interfaces and Refinement. Springer Publishing, New York, NY.
102. Erdobmus, H., and Torchiano, M. 2005. On the Effectiveness of Test-first Approach to Programming. Proceedings of the IEEE Transactions on Software Engineering, 31(1), NRC 47445.
103. Madeyski, L. 2010. Test-Driven Development - An Empirical Evaluation of Agile Practice. Springer Publishing, New York, NY, ISBN 978-3-642-04287-4.
104. Hamilton, M. and Hackler, W. 2000. A Rapid Development Approach for Rapid Prototyping Based on a System that Supports its own Life Cycle. Proceedings, 10th International Workshop on Rapid System Prototyping.
105. Jaakola, H. and Thalheim, B. 2011. Architecture-Driven Modeling Methodologies. In Proceedings of the 2011 Conference on Information Modelling and Knowledge Bases XXII, Anneli Heimbürger et al. (eds), IOS Press.
106. Ramesh, B. 1998. Factors Influencing Requirements Traceability Practice. Communications of the ACM, 41(12):37–44.
107. Lormans, M. and Van Deursen, A. 2006. Can Isi help reconstructing Requirements Traceability in Design and Test? Conference on Software Maintenance and Reengineering, IEEE, Piscataway, NJ, ISBN 0-7695-2536-9.
108. Wagner, S. and Deissenboeck, F. 2007. In 5th Workshop on Software Quality, IEEE Computer Society Press.
109. Davis, A. 1993. Software Requirements: Objects, Functions, and States. Prentice-Hall, Inc., Upper Saddle River, NJ.
110. Navarro, E., Ramos, I., and Perez, J. 2003. Software Requirements for Architectured Systems. In Proceedings of the 11th IEEE International Requirements Engineering Conference, Kyoto, Japan.
111. Luckey, M., Fernandez, D., Baumann, A., and Wagner, S. 2010. Reusing Security Requirements Using an Extended Quality Model. In Workshop SESS at the IEEE International Conference on Software Engineering, Cape Town, South Africa.
112. DeLaurentis, D. 2005. Understanding Transportation as a System of Systems Design Problem. 43rd AIAA Aerospace Sciences Meeting, Reno, Nevada, AIAA-2005-0123.
113. Liu, L. and Yu, E. 2001. From Requirements to Architecture Design – Using Goals and Scenarios. In From Software Requirements to Architectures Workshop (STRAW), Toronto, CAN.
114. Saunders, T. 2005. System-of-Systems Engineering for Air Force Capability Development: Executive Summary and Annotated Brief. US Air Force Publication, Scientific Advisor Board (Air Force), SAB-TR-05-04, Washington, DC.
115. Joshi, C., Trani, A., and Baik, H. 2005. Development of a Decision Support Tool for Planning Rail Systems: an Implementation in TSAM. MS Thesis, Virginia Polytechnic Institute and State University, Blacksburg, VZ.

116. Crowder, J. and Friess, S. 2014. The Agile Manager: Managing for Success. Springer Publishing, New York, NY, ISBN 978-3-319-09017-7.
117. Bettencourt da Cruz, D. and Penzenstadler, B. 2008. Designing, Documenting, and Evaluating Software Architecture Technical Report TUM-I0818, Technische Universität München.
118. Sage, A., and Cuppan, C. 2001. On the Systems Engineering and Management of Systems of Systems and Federations of Systems. Information, Knowledge, Systems Management, Vol. 2, No. 4, pp. 325–345.
119. Parnas, D. 1972. On the Criteria to be used in Decomposing Systems into Modules. Communications of the ACM, 15(12):1053–1058.
120. DeLaurentis, D. A. and Callaway, R. 2004. A System of Systems Perspective for Future Public Policy. Review of Policy Research, Vol. 21, No. 6, pp. 829–837.
121. Edvardsson, J. October 1999. A Survey on Automatic Test Data Generation. Proceedings of the Second Conference on Computer Science and Engineering in Linkoping. CiteSeerX: 10.1.1.20.963.
122. Papoulis, A. 1965. Probability, Random Variables, and Stochastic Processes. McGraw Hill, New York, NY.
123. Ferguson, R. and Korel, G. 1996. The Chaining Approach for Software Test Data Generation. ACM Transactions of Software Engineering and Methodology, 5(1):63–86.
124. Popper, S., Bankes, S., Callaway, R., and DeLaurentis, D. 2004. System-of-Systems Symposium: Report on a Summer Conversation, Potomac Institute for Policy Studies, Arlington, VA.
125. Gold-Bernstein, B., Ruh, W. 2005. Enterprise integration: the Essential Guide to Integration Solutions, Addison Wesley, Boston, MA.
126. Dierolf, D. and Richter, K. 2000. Concurrent Engineering Teams. Technical Report, Institute for Defense Analysis.
127. Fowler, M. 1999. Refactoring - Improving the design of existing code. Addison Wesley Longman, Inc., Boston MA, ISBN 0-201-48567-2.
128. Held, J. 2008. The Modelling of Systems of Systems. PhD Thesis, University of Sydney, Sydney, AU.
129. Carlock, P., and Fenton, R. 2001. System-of-Systems (SoS) Enterprise Systems for Information-Intensive Organizations. Systems Engineering, Vol. 4, No. 4, pp. 242–261.
130. Glinz, M. 2007. On Non-Functional Requirements. International Conference on Requirements Engineering, New Delhi, India.
131. Kotov, V. 1997. Systems-of-Systems as Communicating Structures. Hewlett Packard Computer Systems Laboratory Paper HPL-97-124, pp. 1–15.
132. Crowder, J. 2003a. Using a Large Linguistic Ontology for Network-Based Retrieval of Object-Oriented Components. NSA Technical Paper – CON-SP-0014-2003-03.
133. Crowder, J. 2003c. Agile Business Rule Processing. NSA Technical Paper – CON-SP-0014-2003-06.
134. Crowder, J. 2002. Flexible Object Architecture for the Evolving, Life-like Yielding, Symbiotic Environment (ELYSE). NSA Technical Paper - CON-SP-0014-2002-09.
135. Crowder, J. 1993. Making Change an Integral Component of Advanced Design Methodologies. Academic Press, San Diego, CA.
136. Luskasik, S.J. 1998. Systems, Systems-of-Systems, and the Education of Engineers. Artificial Intelligence for Engineering Design, Analysis, and Manufacturing, Vol. 12, No. 1, pp. 55–60.
137. Taylor, F. 1911. Shop Management, The Principles of Scientific Management. Harper & Row, New York, NY.
138. Menker, L. 2000. Results of the Aeronautical Systems Division Critical Process Team on Integrated Product Development. Technical Report, Wright Patterson AFB.
139. Staszewski, J. 2004. Models of expertise as blueprints cognitive engineering: Applications to land mine detection. Proceedings of the 48th Annual Meeting of the Human Factors and Ergonomics Society, 48, 458–462.

Index

J.A. Crowder et al., *Multidisciplinary Systems Engineering*,
DOI 10.1007/978-3-319-22398-8

MIX
Papier aus verantwortungsvollen Quellen
Paper from responsible sources
FSC® C105338

If you have any concerns about our products,
you can contact us on
ProductSafety@springernature.com

In case Publisher is established outside the EU,
the EU authorized representative is:
Springer Nature Customer Service Center GmbH
Europaplatz 3, 69115 Heidelberg, Germany

Printed by Libri Plureos GmbH
in Hamburg, Germany